polymers,

the environment and sustainable development

6/03

D1264939

polymers,

the environment and sustainable development

Adisa Azapagic, Alan Emsley, Ian Hamerton
University of Surrey, Guildford, UK

Edited by Ian Hamerton

WILEY

St. Louis Community College
at Meramec
Library

Copyright © 2003   John Wiley & Sons Ltd, The Atrium, Southern Gate, Chichester,
West Sussex PO19 8SQ, England

Telephone (+44) 1243 779777

Email (for orders and customer service enquiries): cs-books@wiley.co.uk
Visit our Home Page on www.wileyeurope.com or www.wiley.com

All Rights Reserved. No part of this publication may be reproduced, stored in a retrieval system or transmitted in
any form or by any means, electronic, mechanical, photocopying, recording, scanning or otherwise, except under
the terms of the Copyright, Designs and Patents Act 1988 or under the terms of a licence issued by the Copyright
Licensing Agency Ltd, 90 Tottenham Court Road, London W1T 4LP, UK, without the permission in writing of the
Publisher. Requests to the Publisher should be addressed to the Permissions Department, John Wiley & Sons Ltd,
The Atrium, Southern Gate, Chichester, West Sussex PO19 8SQ, England, or emailed to permreq@wiley.co.uk, or
faxed to (+44) 1243 770620.

This publication is designed to provide accurate and authoritative information in regard to the subject matter
covered. It is sold on the understanding that the Publisher is not engaged in rendering professional services. If
professional advice or other expert assistance is required, the services of a competent professional should be
sought.

Other Wiley Editorial Offices

John Wiley & Sons Inc., 111 River Street, Hoboken, NJ 07030, USA

Jossey-Bass, 989 Market Street, San Francisco, CA 94103-1741, USA

Wiley-VCH Verlag GmbH, Boschstr. 12, D-69469 Weinheim, Germany

John Wiley & Sons Australia Ltd, 33 Park Road, Milton, Queensland 4064, Australia

John Wiley & Sons (Asia) Pte Ltd, 2 Clementi Loop #02-01, Jin Xing Distripark, Singapore 129809

John Wiley & Sons Canada Ltd, 22 Worcester Road, Etobicoke, Ontario, Canada M9W 1L1

Wiley also publishes its books in a variety of electronic formats. Some content that appears
in print may not be available in electronic books.

Library of Congress Cataloging-in-Publication Data

Azapagic, Adisa.
   Polymers : the environment and sustainable development / Adisa Azapagic, Alan
Emsley, Ian Hamerton ; edited by Ian Hamerton.
      p. cm.
   Includes bibliographical references and index.
   ISBN 0-471-87740-9 (cloth : alk. paper) – ISBN 0-471-87741-7 (pbk. : alk. paper)
   1. Plastics – Recycling. 2. Sustainable development. I. Emsley, Alan. II. Hamerton, Ian.
III. Title.

TP1175.R43 A9 2003
668.4'192 – dc21   2002028088

British Library Cataloguing in Publication Data

A catalogue record for this book is available from the British Library

ISBN 0-471-87740-9
ISBN 0-471-87741-7

Typeset in 10/12pt Rotisanserif by Laserwords Private Limited, Chennai, India
Printed and bound in Great Britain by Antony Rowe Ltd, Chippenham, Wiltshire
This book is printed on acid-free paper responsibly manufactured from sustainable forestry
in which at least two trees are planted for each one used for paper production.

FRONT COVER IMAGE
Jason and Madea; Artist: John William Waterhouse; Painting Date: 1907; Medium: Oil on canvas; Size:
134 × 107 cm; Location: Unknown.
About the painting: During the adventure of the Argonauts, Jason put ashore at Colchis where he met Medea, the
daughter of Aeetes, and was bewitched by her beauty. Aeetes, the King of Colchis, obstructed Jason's quest for
the golden fleece by setting him an impossible task, but Medea, being in love with him, helped him perform it by
magic and escaped with him to Greece. Overcome by wrath, Aeetes pursued her and, in an effort to delay his
advances, Medea murdered her brother, strewing his mutilated limbs in her father's path. On their arrival at
Iolcos, Medea rejuvenated Jason's father Aeson by boiling him with magic herbs but her evil trickery forced them
to flee to Corinth, where Jason deserted her for Glauce. Medea took revenge by slaughtering their children and
poisoning her rival.

about the authors

Adisa Azapagic was born and educated in Tuzla, Bosnia. She holds Dipl.-Ing. and MSc in Environmental Chemical Engineering from the University of Tuzla, and a PhD in Environmental Systems Analysis from the University of Surrey. She worked for eight years at the University of Surrey as a Senior Teaching and Research Assistant, before moving to the UK in 1992. After spending a year at the University of Leeds as a British Council Senior Research Fellow, she moved to the University of Surrey, where she is currently based. She became Lecturer in the Department of Chemical and Process Engineering in 1996 and was promoted to Reader in 2000. Her main research interests and expertise are in system modelling and optimisation, clean technology, life cycle thinking, industrial ecology and sustainable development. She is also interested in engineering education for sustainable development. She has over 100 publications in these areas of interest. She has been a holder of various fellowships and awards, of which the latest come from the Leverhulme Trust and the Royal Academy of Engineering.

Alan Emsley was born in Preston Lancashire in 1946 and educated in the grammar school system in Preston and in Knaresborough, Yorkshire. He holds a BSc(Hons) and Ph.D in Chemistry from the University of Edinburgh and was a research officer at the Central Electricity Research Laboratories, Leatherhead, Surrey (later National Power Laboratories) for 23 years, before joining the University of Surrey as a Senior research Fellow in 1992. His research interests are in the use of chemical kinetics to predict the life expectancy of materials and span all aspects of the degradation of materials from oxidation of metals and carbon dusting in high pressure gas environments to the effects of degradation of polymeric structures in industrial applications. He is an international expert on the degradation of cellulose, with particular interest in its effects on the life expectancy of electrical transformers, where it is used as the primary source of insulation on the high voltage windings. His interests in sustainable development arise from studies of the effects of the environment on polymer degradation and the issues around polymer life expectancy.

Ian Hamerton was born in Southampton, Hampshire in 1964 and attended the local comprehensive (Hightown School) and Itchen Sixth Form College. He holds a B.Sc. (Hons.) and Ph.D. in Chemistry from the University of Surrey. Following a period as a Research Fellow (a joint post with the Royal Aerospace Establishment, Farnborough), he joined the University of Surrey as a Lecturer in Organic Chemistry in 1991 and was promoted to Senior Lecturer in 1996. A Fellow of the Royal Society of Chemistry and an Associate Member of the Institute of Materials, his research interests span many aspects of pure and applied polymer chemistry. He is an international expert in the field of high performance polymers for structural composites, with particular interests in cyanate esters, polyimides and epoxy resins, having published over 100 articles (including several patents and an edited book) in the area. His interest in sustainable development has arisen from the perspective of green chemistry: the simulation of polymer structure and properties, the study of electron- and proton-beam cure of polymers and the preparation of structural composites from renewable resources. Married to Sally, with a young son (Archie), Ian's interests outside the subject include collecting antique furniture and decorative objects produced by the Arts and Crafts Movement and studying art of the Pre-Raphaelite Brotherhood. Ian is an active committee member of the Society for the Arts and Craft Movement in Surrey and recently lived in a lodge designed by the architect/designer C. F. A. Voysey. In addition to his polymer research he is now involved in two writing projects featuring the designer W. A. S. Benson and Compton pottery.

preface

Shakespeare conceived the ultimate life cycle, literally from cradle-to-grave. In *As You Like It*, act II part 7, Jacques says:

'All the world's a stage and all the men and women merely players. They have their exits and entrances and one man in his time plays many parts, his acts being seven ages.'

His seven ages were the puking infant, the whining schoolboy, the sighing lover, the soldier full of oaths, the round-bellied justice, the lean and slippered pantaloon and then the final act (second childhood and mere oblivion) *'sans teeth, sans eyes, sans taste, sans everything'*.

Using this metaphor in relation to polymers, we can paraphrase Jacques' seven ages in terms of the seven stages in the life cycle of a polymer as: production, use, waste generation, post-consumer waste recovery, re-use, recycling, and landfilling. Later on we will extend this analogy into a 'life guide' to polymeric materials.

The issue of use and management of resources is central to creating sustainable societies. It is also an issue that is very difficult to address: although we all realise that we are consuming too much of the earth's resources, we are not prepared to compromise our life styles now for a more sustainable living in a distant future. Therefore, a question that is often asked by many is: 'How can we be more sustainable in our use of resources and management of waste now so that we do not compromise the ability of the future generations to enjoy the benefits of these resources?'

Polymeric materials and products, which are ubiquitous in our everyday life, have a potential to generate a significant impact on the environment throughout their life cycle, including depletion of finite, non-renewable, resources and generation of solid waste. Therefore, it is essential that we identify more sustainable ways of using and managing polymers. This is not a trivial task and at the very least it requires finding out how to:

- optimise production and use of polymers in order to minimise resource use, waste generation and the post-consumer waste recovery problem;
- devise new and efficient ways to re-use and recycle used polymeric materials in the manufacture of new products;
- when re-use and recycling options are exhausted, recover the feedstock energy stored in polymeric materials, in order ultimately to;
- eliminate the need for the use of landfill for the dumping of waste.

This book takes on a challenging task in trying to provide some answers to these questions. By adopting a life cycle approach, we follow the polymeric materials from 'cradle to grave' in an attempt to identify the most sustainable options for the production, use and management of polymeric resources.

As a result, this book could be viewed as a 'life guide' to polymeric materials, with the first chapter providing an introduction and each subsequent chapter concentrating on one stage in the life cycle of polymers. Therefore, the structure of the book is as follows:

Life Guide (Chapter 1):
The Environment and sustainable development: an integrated strategy for polymers.

Facts of Life (Chapter 2):
Polymers in everyday use: principles, properties and environmental effects.

First Life (Chapter 3):
Feeding the waste streams: sources of polymers in the environment.

Second Life and Beyond (Chapter 4):
Managing polymer waste: technologies for separation and recycling.

Life Forces (Chapter 5):
Drivers and barriers for polymer recycling: social, legal and economic factors.

Life After Life (Chapter 6)
Design for the environment: the life cycle approach.

Better Life (Chapter 7):
Environmental impacts of recycling.

Life Hereafter (Chapter 8):
Future directions: towards sustainable technologies.

This book is primarily aimed at advanced undergraduate or postgraduate students in polymer, materials or environmental science, but it will also be of interest to engineers and other scientists (including social sciences) who want to learn more about polymers, their impact on the environment and the relevance for sustainable development.

The multidisciplinary nature of the book means that different readers will come to it from different backgrounds. Therefore, the book has been written in such a way that each chapter builds on the previous one and yet is largely self-contained and can be read in isolation, if you already have the background knowledge.

Thus Chapter 1, which provides a general introduction to sustainable development and puts the issue of polymers in the context, could be of interest to a reader of any background. Chapters 2 and 3 provide scientific facts on polymer properties and production processes, as well as an overview of the sources of polymeric materials in our waste streams. Hence, this chapter may be attractive to a **chemical engineer** who may wish to refresh his/her knowledge of polymer structures and production processes, whereas a **polymer chemist** or **material scientist** may be able to skip these and go straight into Chapter 4, where we examine various technological options for recycling. This is followed in Chapter 5 by a discussion on why it is difficult to increase the current recycling rates and what socio-economic factors and legal instruments could facilitate a more sustainable waste management. A **social scientist** may find that he/she is already familiar with the socio-economic factors but wants to know more about the legislative drivers for recycling discussed in Chapter 1. Anyone not familiar with life cycle thinking and related tools, such as life cycle assessment (LCA) and life cycle product design (LCPD) will find an in-depth discussion in Chapter 6 followed by some practical polymer-related applications in Chapter 7. These chapters may in particular be relevant to an **environmental scientist**, **product designer** or an **engineer**. Finally, in Chapter 8 we examine some of the technologies currently under development, which may provide us with the means for a more sustainable design, use and re-use of polymer resources. Both scientists and engineers should find this chapter of interest. Concluding, we argue that, although technological solutions are important, they alone are unlikely to provide all answers to the problem and participation of the whole of society is necessary for a more sustainable resource management.

acknowledgments

The authors are grateful to a number of people for their help with the book. We are in particular indebted to Rosalind Malcolm, University Director in Law at the University of Surrey and Barrister in Environmental and Property Law in Field Court Chambers, Gray's Inn, who contributed the sections on environmental legislation in Chapter 1 and also wrote Appendix 1. We also thank Lesley Charlton (of the TCI Science Education Unit) for supplying information about the Teesside TEC collaborative learning project.

Adisa Azapagic would like to acknowledge her colleagues at the University of Surrey for their input and contribution over the years to some of the work presented in Chapters 6 and 7, in particular to Roland Clift, Warren Mellor, Gary Stevens and Elizabeth Wright. Special thanks are due to my husband and a colleague, Slobodan Perdan, who endured endless 'philosophical' and 'practical' discussions on the various aspects of the book and provided the much-needed support for the completion of my part of this project.

Alan Emsley would like to thank his long-suffering wife for "time-off" from domestic duties, evenings and weekends for researching and writing time to keep the project moving "out-of-hours".

Ian Hamerton would like to express his gratitude to Dr Brendan Howlin of the University of Surrey for his constructive comments associated with the preparation of the molecular modelling section and to Paul Klewpatinond (also of Surrey), who kindly prepared a diagram from data arising from his PhD research. I am particularly grateful to the many members of staff at Wiley who have become associated with this project during the last three years. Particular mention should be made of the assistance offered by Jenny Cossham, Lynette James, Chris Pote, Helen McPherson, Katya Vines, Leigh Murray and Andrew Slade. The collective efforts of these stalwarts and my co-authors have transformed my original concept of a textbook covering the impact of the use of polymers on the environment and the potential of recycling to address this increasing problem, to include a broader discussion of sustainable development. My special thanks are due to my wife Sally, who has listened patiently, and I dedicate my work to her and to our beautiful young son, Archie.

photo credits

CHAPTER OPENING IMAGES

Chapter 1 – Pandora, 1871 (oil on canvas) by Dante Gabriel Rossetti (1828–82) Private Collection/Bridgeman Art Library

Chapter 2 – The Doom Fulfilled (Perseus Slaying the Sea Serpent) c.1876 (gouache on paper) by Sir Edward Burne-Jones (1833–98) Southampton City Art Gallery, Hampshire, UK/Bridgeman Art Library

Chapter 3 – J.W. Waterhouse, Great Britain, 1849–1917, Circe Invidiosa, 1892, Primrose Hill Studios, London, oil on canvas, 180.7 × 87.4 cm, Art Gallery of South Australia, Adelaide, South Australian Government Grant 1892

Chapter 4 – The Wheel of Fortune, 1875–83 (oil on canvas) by Sir Edward Burne-Jones (1833–98) Musee d'Orsay, Paris, France/Bridgeman Art Library

Chapter 5 – Love and the Pilgrim, tapestry designed by the artist and woven by Morris & Co., 1909 (textile) by Sir Edward Burne-Jones (1833–98) Birmingham Museums and Art Gallery/Bridgeman Art Library

Chapter 6 – The Magic Circle, 1886 (oil on canvas) by John William Waterhouse (1849–1917) Christie's Images, London, UK/Bridgeman Art Library

Chapter 7 – The Mirror of Venus, 1870–76 (oil on canvas) by Sir Edward Burne-Jones (1833–98) Museu Calouste Gulbenkian, Lisbon, Portugal/Bridgeman Art Library

Chapter 8 – The Damsel of the Sanct Grael, 1874 (print on canvas) by Dante Gabriel Rossetti (1828–82). Illusions Gallery, USA.

PHOTOS AND ILLUSTRATIONS

Chapter 2
Photo of Polymerisation Plant – Copyright Digital Vision

Chapter 3
Earth Image – Copyright Digital Vision
Municipal Solid Waste – Copyright Photo Researchers, Inc.
Car Seats – Copyright Corbis Digital Stock
House Under Construction – Copyright Photodisc, Inc.
Packaging Bales – Copyright Corbis Digital Stock
Crop Harvest – Copyright Photodisc, Inc.
PCBs processing – Copyright Digital Vision

Chapter 5
Waste clean up – Copyright Photodisc, Inc.
Education – Photograph by Linda Westmore, and courtesy of the University of Surrey, UK
Classroom – Photograph by Linda Westmore. Courtesy of the University of Surrey, UK.

the environment and sustainable development: an integrated strategy for polymers

Chapter 1 – Pandora (DG Rossetti, 1871).
In Greek mythology, Pandora was the first woman, fashioned from clay by Hephaestus at Zeus' command. Pandora was made a gift of a box, containing all the ills and diseases, by Zeus to present to her future husband and thus destroy Prometheus' creation of man. Sadly, the box was opened and the ills and diseases unleashed into the world leaving only hope lingering at the bottom of the box, to console mankind – a fitting start to our examination of the environmental impacts of polymers and the ultimate hope of achieving sustainable development.

1.1 INTRODUCTION TO SUSTAINABLE DEVELOPMENT

'The existing pattern of resource use will lead to a collapse of the world system within the next century'. These were the words that hit the headlines when the world was shaken by the first oil crisis in 1973. This viewpoint, advocated in *The Limits to Growth*[1], dominated thinking throughout the 1970s and much of the 1980s and led to a wide acceptance of the depletion of resources as a central environmental, economic and political issue. It was based on the premise that natural resources, particularly oil, were about to run out. This pessimistic prediction has, however, proved to be false and the collapse of oil prices in 1986 marked the end of 'the era of resource scarcity'. New concerns over the future of the global environment then started to emerge.

One of these was the keen sense of human vulnerability to environmental changes. It soon became apparent that a unifying approach to concerns over the environment, economic development and the quality of life was necessary if human (and other) life was to be sustained for an indefinite period in the future. This approach, which developed slowly from the early 1980s and is now widely accepted, is generally referred to as Sustainable Development.

The idea of sustainable development was first used in the *World Conservation Strategy* report[2] by the International Union for the Conservation of Nature, published in 1980. It was followed in 1983 by the Brandt Commission's *Common Crisis*[3] which in effect was the forerunner of, and in many ways formed the basis to, the report *Our Common Future*[4], published in 1987 by the World Commission on Environment and Development. This publication, also known as the Brundtland report, set the benchmark for all future discussions of sustainable development and gave the most commonly used, working definition of sustainable development as that which 'meets the needs of the present without compromising the ability of future generations to meet their own needs'.

In essence, the Brundtland report called for policies which foster economic growth but also satisfy the needs of people and improve quality of life without depleting the environment. This vision of sustainable development required a different attitude to economic development, in which the quantity of growth is replaced by the quality of growth.

The Brundtland report prompted numerous actions at both national and international levels, which called on governments, local authorities, businesses and consumers to define and adopt strategies for sustainable development. One of the most notable of these activities, instigated as a direct consequence of the emergence of the concept of sustainable development, was the Earth Summit held in Rio de Janeiro in June 1992. The Summit was attended by 120 world leaders and representatives from over 150 countries and adopted a comprehensive action plan known as Agenda 21[5], for the pursuit of sustainable development.

In response to the Agenda, many governments and organisations started developing their own plans of action and setting out strategies for sustainable development. Countries such as Sweden, Canada, Germany and the UK have already started working towards their own sustainability targets and, more recently, the EU sustainable development strategy[6] has also been adopted.

1.2 SUSTAINABLE DEVELOPMENT ISSUES

Sustainable development may be regarded as the progressive and balanced achievement of sustained economic development, improved social equity and environmental quality[7]. This concept has both spatial and temporal dimensions as it must satisfy these three goals equally across the globe for both present and future generations. Although holistic in concept, sustainable development comprises three individual components (society, environment and economy) and the goals of sustainable development can only be achieved if all three components can be satisfied simultaneously (see Figure 1.1). For this to happen, a number of global and local problems need to be addressed.

One major issue is global inequity and widespread poverty: 20 % (1.2 billion) of the world's population receives nearly 83 % of total world income. There are significant links between poverty and the environmental quality and much of the environmental degradation we see in the developing world arises as a result of people seeking basic essentials of life: food, water, *etc*. On the other hand, environmental problems are a significant cause of poverty and generally hit the poor hardest, *e.g.* a quarter of all diseases are found in developing countries. One of the main causes of environmental degradation, however, is unsustainable development by the rich. The 'big seven', *i.e.* USA, Japan, Germany, Canada, France, Italy and the UK, make up less than 12 % of the world's population, but consume between 55 and 65 % of world resources. If the rest of the world continued to consume the energy resources as the UK does today, we would need eight and a half planets to sustain current global consumption in 2050 (see Figure 1.2). The patterns of consumption and distribution of resources cannot be sustained if, as currently predicted, the world population grows to 10 billion by the end of the 21st century.

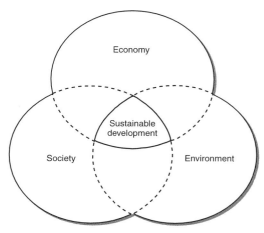

Figure 1.1 The three components of sustainable development

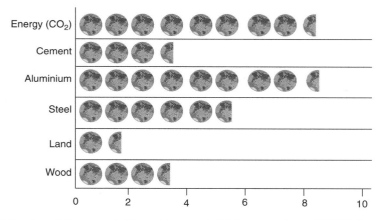

Figure 1.2 Number of planets needed to sustain current global consumption in 2050 if all countries consumed as the UK does today[8]. (Key facts for resource consumption: 12 % of the world's population consume:

- 43 % of the world's fossil fuel production;
- 64 % of the world's paper;
- 55–60 % of all the aluminium, copper, lead, nickel and tin). Reproduced with permission from McLaren *et al.* (1998). Tomorrow's World: Britain's share in a Sustainable Future. Copyright. Friends of the Earth/Earthscan.

Key Facts
- 20 % of the world population receives 83 % of the total income.
- 12 % of the world population consumes 55–65 % of world resources.
- Europe generates some 2.6 billion tonnes of waste a year.

Coupled with other global environmental problems such as climate change and loss of biodiversity, there are clear indications that we are now exceeding the 'carrying capacity' of the environment. This is exacerbated by local or regional issues, such as air pollution and generation of solid waste. For example, some 2.6 billion tonnes of industrial, agricultural and domestic waste is generated each year in Europe alone. The decreasing capacity of landfills and their recognised impact on the environment give waste management a high priority at the local and regional levels.

To enable the move towards sustainability on the practical level, it is first necessary to understand these causes of unsustainability, then to identify more sustainable options and finally to determine how they may be implemented. In doing so, it is paramount that problems and solutions are analysed by adopting more holistic, life cycle thinking. This requires a paradigm shift from the current, fractured view of the environment, with the emphasis on one stage of the life cycle (*e.g.* the production process), to a whole life cycle approach, which examines the consequences of human activities on the environment from 'cradle' (extraction of resources) to 'grave' (disposal of waste)[9,10].

In this book, we adopt such an approach in an attempt to examine the options and contribute towards the practice of sustainable development by addressing two important areas: resource use and waste management. We concentrate on polymeric materials and products, ubiquitous in our everyday life, to try and understand what drives and limits their production, use, re-use and recycling. We will consider a wide range of polymers, but will mainly concentrate on plastic* materials, *i.e.* thermoplastics and thermosets, because they constitute the majority of the market. The conceptual approach adopted in the book is illustrated in Figure 1.3, which shows a 'life guide' for polymers with a number of different lives (or cascades of uses) and the associated life cycle stages. The guide through the chapters is also shown in the figure. We particularly concentrate on post-consumer waste management and examine the influencing technical, legislative, environmental, economic and social factors with the aim of identifying more sustainable options for polymer re-use and recycling.

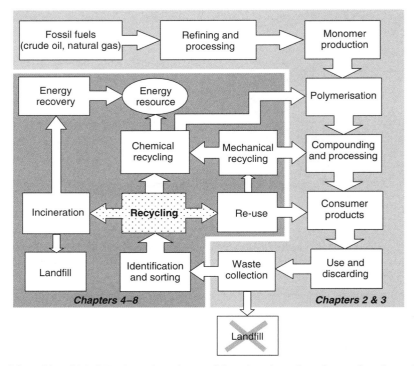

Figure 1.3 A 'life guide': following polymeric materials and products through cascades of uses from 'cradle to grave' (note that both energy and materials are consumed in every life cycle stage)

Before looking into these issues in detail in the chapters that follow, we continue here to examine why polymers may be an issue for sustainable development.

---

*We use the term 'polymer' as a chemical term to describe a macromolecule and the term 'plastics' as a generalisation which covers all polymeric materials but, strictly speaking, 'plastic' defines the stress/strain behaviour of the material and should really only be applied to thermoplastics and thermosets (see Chapter 2).

1.3 POLYMERS: AN ISSUE FOR SUSTAINABILITY

The emergence of the concept of sustainable development has once again made fossil fuels an issue, because it is clear that reserves will run out on time scales relevant to sustainable development, although perhaps not as soon as was predicted in the 1970s. However, scarcity of resources is not the only issue to be considered, since burning fossil fuels affects climate change and it is now widely accepted that the millions of tonnes of CO_2 produced each year by burning fossil fuels are one of the main causes of global warming. We must therefore rethink our use of such fuels and general consumption patterns into a more sustainable model.

Most synthetic polymers are derived from fossil fuels, *i.e.* from naphtha or natural gas (see Figure 1.3), which puts them immediately into the environmental 'spotlight'. Consumption of fossil fuels and the associated environmental damage have made polymeric materials and products a focus of much attention by various environmental and government groups (see Figure 1.2). They have argued that polymers use material and energy resources, which are then lost when the polymers are disposed of, usually in landfill. The production process itself also results in a loss of 'feedstock' energy. For example, the production of 1 tonne of high density polyethylene (HPDE) loses 17.9 GJ of the 71.4 GJ of calorific value in the naphtha feedstock. Put another way, some 40 % of the energy of the original crude oil is lost during processing[11].

However, the consumption of material and energy resources is not the only issue surrounding polymeric materials and products. Because of their widespread use and our 'linear' consumption patterns (in which materials and products are used only once and then discarded), polymers also contribute to an ever-increasing amount of solid waste. Since the 1930s, the total world production and consumption of polymers have risen rapidly to reach figures in excess of 100 million tonnes in 1995, about a quarter of which was produced in Europe. The types of material involved include plastic products (made from both thermoplastics and thermosets), fibres (*e.g.* textiles), elastomers, coatings and adhesives. In Western Europe around 45, mainly multinational companies, produce the basic polymer, which is sold to around 30 000 small- and medium-sized companies. These, in turn, convert the polymer into products for use in many sectors, for example, packaging, automotive parts and electronic equipment. Since 40 % of plastics are used for packaging, it is not surprising that this product category has attracted most attention from policy makers and environmentalists. For example, the total plastics consumption in Western Europe in 1999 was 33.5 million tonnes or 84 kg of plastics *per* person[12], 19 million tonnes of which were available for collection as waste, with the rest remaining in use. Because packaging has a much shorter life than, for instance, plastics used in the construction or automotive industry, it reaches the waste stream much more quickly, which explains the fact that 70 % (or 13 million of tonnes) of the total plastics waste that appeared in the same year was packaging.

On average, polymers account for 7–8 % by weight and 20 % by volume of municipal solid waste in Europe and elsewhere. Of that, still relatively little is recycled. For example, in Western Europe only 6 million tonnes or 30 % of the total post-consumer waste were recycled in 1999[12], with the rest going to landfill. Similar trends are found in other parts of the world. Not only does this practice waste valuable resources, but it also has negative impacts on the environment. Very few polymers are biodegradable so that, once in a landfill, they will remain there occupying space for a long time; according to some estimates, up to 200 years for some polymers. However, some of the additives used to improve polymer properties can leach from a landfill to contaminate the water table; or in poorly managed landfills burning of plastic waste can generate toxic substances and cause air pollution.

Furthermore, as we all know, not all polymer waste reaches the landfill; much of the waste also remains abandoned and scattered in the streets of our cities and towns, as well as in the countryside, affecting the aesthetic aspects of life.

Key Facts
- 40 % of the energy of crude oil is lost during the manufacture of high density polyethylene.
- World consumption of polymers reached 100 million tonnes in 1995.
- 40 % of plastics are used for packaging, 84 kg *per* person *per annum* in Europe in 1999.
- Polymers account for 7–8 % by weight of post-consumer plastic waste.
- 70 % of post-consumer waste in Europe went to landfill in 1999.

It is thus apparent that continuing with the same 'make – use – discard' practice is unsustainable because it leads to generation of waste, loss of resources (material and economic), environmental damage and also raises social concerns. Hence, we need to identify more sustainable practices for polymeric materials and products. The following section gives an overview of the options available, which are then considered in more detail later in the book.

1.4 INTEGRATED RESOURCE AND WASTE MANAGEMENT

The fact that only 4 % of the world's oil reserves are used in the manufacture of polymers is sometimes used as an argument that they do not contribute much to the degradation of the environment, but 4 % still represents a valuable resource. Furthermore there are other issues to consider, such as the generation of (long-lived) solid waste and pollution associated with polymeric materials and products. Hence addressing the problem of polymers in the environment remains an important goal.

The use of resources and management of waste in a more sustainable fashion cannot be achieved in any single way. However efficiently we use resources, the laws of thermodynamics teach us that some waste will always be generated. This, coupled with increasing consumption and the fact that it is difficult to persuade people to change their life styles, requires an integrated resource and waste management strategy. The waste management hierarchy shown in Figure 1.4 involves following the options of reduction, re-use, recycling, incineration and landfill.

Key Facts
● 4 % of the world's oil reserves are used in the manufacture of polymers.
● Waste management involves reduction, re-use, recycling, incineration and finally landfill as some waste is unavoidable.

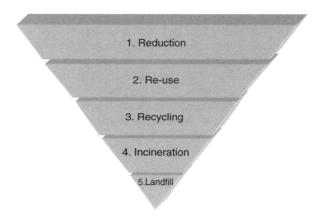

Figure 1.4 Resource and waste management hierarchy in a decreasing order of desirability

The most desirable option in this hierarchy is reduction of resource use, which also leads to a reduction in the generation of waste. The next two options are aimed at turning waste back into resources through re-use and recycling of materials, leading to conservation of natural resources and reduction of other environmental damage. Thus, adopting a 'more with less' approach maximises benefits from products and services, uses the minimum amount of resources and rejects the least amount of waste or emissions to the environment. In essence, the production of waste is seen as a demonstration of the inefficient management of resources. This is very much in harmony with the laws of nature, where there is no such thing as waste. All biological systems are interconnected and what is waste for one system is a valuable resource for another. This concept is also known as the industrial ecology of materials and further reading on the topic is provided by Graedel and Alenby[13].

The last two options in the hierarchy are incineration (without energy recovery) and landfill. Because they both waste valuable resources, with incineration also contributing to air pollution, they are not considered to be sustainable options. However, it should be borne in mind that, even with the first three options fully implemented, some waste is still unavoidable and has to be disposed of by either incineration or landfill.

polymeric materials and products.

1.4.1 Reduction

The aim for the future must be to design products so as to minimise the use of materials and energy in the manufacturing and use stages and minimise waste and emissions to the environment, a concept known as dematerialisation. There are various ways to dematerialise our economy and the reader is referred to the book by Jackson[14] for a detailed exposition on the subject.

Various approaches have been developed to facilitate reduction of resource use and they are known collectively as Design For the Environment (DFE). They apply life cycle thinking and use Life Cycle Assessment (LCA) as a tool to enable the design of products, which not only minimise the use of resources but are also easy to disassemble, re-use and recycle. LCA follows a product or an activity from 'cradle to grave', *i.e.* from the extraction of raw materials, the production and use, the re-use and recycling options to the final disposal. It quantifies environmental impacts associated with each of these stages to provide a full picture of the impact of an activity on the environment. Taking such a holistic approach to design ensures that environmental impacts are not merely shifted upstream or downstream in the supply chain, thus giving a true picture of the total consequences of an activity on the environment. This approach also enables innovation and technological improvements by identifying the 'hot spots' or major concerns that need addressing. The general principles of DFE and LCA and their application to polymers are discussed in detail later in this book.

DFE principles have already been applied to polymers, particularly to plastic products, which has led to an average decrease in the weight of plastics packaging of approximately 28 % in the last 10 years[15]. Dematerialisation has saved more than 1.8 million tonnes of plastics (over the 10 years), which is greater than the total volume of mechanically recycled post-user plastics for all applications in 1998 (1.6 million tonnes). The additional benefits of reducing the weight of products are reduced environmental impacts and costs associated with their transportation.

Finally, it is also important to mention that the use of plastics compared to other alternatives can save materials and energy. For instance, in Western Europe the automotive sector uses 1.7 million tonnes of plastics a year, made from the equivalent of 3.25 million tonnes of oil. However, it is estimated that 12 million tonnes of oil are saved each year through fuel efficiencies, because vehicle components manufactured from plastics are lighter than metal equivalents, leading to a saving in CO_2 emissions of 30 million tonnes a year[15]. In the same reference, it is claimed that the use of nonplastic packaging would increase overall packaging consumption by 291 % by weight, with an increase in manufacturing energy of 108 % and volume of waste of 158 %. Another estimate also shows energy savings in the use of plastic packaging compared to the alternatives: the projected savings made each year are enough to power a city of 1 million homes for roughly 3.5 years[16]. However, these results have to be interpreted with care as they refer to the use stage only and do not include other life cycle stages, such as extraction and processing of raw materials, manufacture of packaging and post-consumer waste management.

Reduction of resource use through better design is not sufficient on its own, unless it is accompanied by more sustainable consumption. Although we have seen substantial dematerialisation in many parts of the economy in the past years, the benefits have hardly been obvious and the main reason for that is a constant increase in consumption. One typical example is the use of mobile telephones. Although their weight and the amount of materials used for the manufacture have gone down in the past few years[17], by at least a factor of 10, the market has expanded so much that the resources used for their manufacture have in fact increased. In 1997 alone, 100 million mobile telephones were

Key Facts
- Design For the Environment reduces the use of resources and facilitates re-use and recycling.
- Life Cycle Assessment quantifies the environmental impact of a product 'from cradle to grave'.
- Dematerialisation has saved 1.8 million tonnes of plastics over the last 10 years.
- 12 million tonnes of oil are saved each year in transportation, because plastic components in vehicles are lighter than the metal equivalents.

Key Facts
● Moving from
profligate
consumption to
prudent use of
resources requires a
change in lifestyle.
● Re-use is impeded
by dispersion of
products in the
marketplace, ease
of disassembly and
reluctance of
consumers to
accept products
that are not brand
new.
● Re-use is
ultimately limited
by degradation of
the material's
properties.

sold world-wide; sales in Western Europe, for example, grew by a factor of 18 in the period 1991–1997[17]. Given their fairly short lifetime and obsolescence due to technological developments, they generate annually a large amount of waste (1080 tonnes in Western Europe in 1997[17]). Add to that the fact that, on a life cycle basis, the industrial operations for their manufacture generate solid waste roughly 200 times the weight of the telephone itself[18], and you can see the scope of the problem.

Moving away from profligate consumption towards more prudent use of resources will inevitably require some changes to our life styles. At present, few people are prepared to accept or do that. This therefore remains an option that has to be viewed as a long-term target. In the meantime, we have to pursue the other, short- to medium-term objectives as defined by the integrated strategy for resource and waste management. Hence, the next option to examine is resource re-use.

1.4.2 Re-use

One of the reasons for the widespread use of polymers is their versatility and, in particular, their strength and durability. The same properties can be exploited for their re-use in further applications and some manufacturers are already reclaiming plastic parts from their used products to re-use them in new products. One of the typical examples is Xerox who re-use plastic (and other) parts from old photocopiers in the manufacture of new machines.

Although this remanufacturing process is gaining wider acceptance, particularly among the manufacturers of cars and electrical and electronic equipment, there are at least three obstacles to its becoming a normal practice. Firstly, the manufacturer must be able to recover their products from customers. Since most products are sold on an individual basis, they become highly dispersed in the marketplace and it is impossible for the original manufacturers to keep track and reclaim them back from customers. In many cases, the customer does not necessarily want to own an artefact, but wants the service it delivers, so one way to overcome the recovery problem is to switch from sale of articles to leasing. In this way, the manufacturer sells a service and retains ownership of the product. It then becomes a simple exercise to recover it at the end of its life[19]. This is the approach taken by Xerox, who lease photocopiers and, in effect, sell the photocopying service rather than the machines.

Secondly, the re-use of parts from products recovered at the end of their useful life also depends on the design of a product, i.e. how easy it is to disassemble into its constituent parts. Complex products are particularly difficult to dismantle and the parts can be damaged during the process, making them unusable. Many electronic products are designed this way, including mobile telephones and TV equipment. Here, a DFE approach to manufacture would facilitate dismantling/disassembly and re-use.

The third obstacle to re-use is customer perception: many people are still reluctant to accept products which are not brand new, because they believe that the performance of remanufactured products is inferior to that of new products. Reducing the price of remanufactured photocopiers is one of the ways in which Xerox try to encourage their customers to lease these rather than brand new machines. On the other hand, consumers are prepared to re-use individual polymer products in their households. For example, many people are routinely re-using plastic shopping bags, containers and water bottles so that these products effectively stay longer in the use phase than originally intended by the manufacturers.

However, the number of re-use cycles is limited and eventually the properties of polymers start to deteriorate to the point when they can no longer be used without further processing. This brings us to the third option in the resource and waste management hierarchy, i.e. recycling.

At the end of their first life cycle, or perhaps after being re-used several times, polymers can be recycled to yield new polymeric materials or products. The following options exist to take further advantage of the valuable material and/or energy resources still stored in them:

- mechanical recycling,
- chemical recycling,
- energy recovery.

The mechanical and chemical recycling options are collectively termed 'material recycling' because they recycle plastics back into usable materials or fuels respectively, as distinct from the third option that recovers energy.

As mentioned earlier, in this book we are particularly concerned with the recycling options for polymers so they will be discussed at length later. Here, however, we give just a brief overview of each option, before continuing on to talk about waste management policies that influence recycling.

Mechanical Recycling

Mechanical recycling uses physical and mechanical means, such as grinding, heating and extruding to process waste plastics into new products. It requires clean and homogeneous waste, which means that plastics have to be sorted by type and separated before they can be incorporated in virgin polymers of the same type, or used on their own. The availability of homogeneous waste streams of known characteristics is thus a key criterion for successful recycling.

Chemical Recycling

This is another form of material recycling, which is particularly well suited to mixed plastics waste. It uses chemical processes to break the polymers down into their chemical constituents and convert them into useful products, such as basic chemicals and/or monomers for new plastics or fuels. As in mechanical recycling, some pretreatment of plastic waste is required to meet the specification of the recycling process.

Energy Recovery

If material recycling is not viable or after certain products have been removed from the waste stream for mechanical recycling, the high calorific value of plastic waste can be recovered as energy[20]. Energy recovery can be achieved by direct incineration, *e.g.* in municipal waste incinerators to generate heat and electricity; or waste polymers can be used directly in production processes to replace other fuels (*e.g.* in cement kilns) or for power generation.

At present, the majority of post-consumer waste is recycled as energy, followed by mechanical recycling and, at much lower rates, by chemical recycling. The rates of recycling are different in different countries but overall they are still very low. In Western Europe only 30 % of polymer waste is recycled and the rest goes to landfill. However, there is an indication that the recycling rates may be increasing. For example, according to some estimates, mechanical recycling in Western Europe has the potential to double in the period 1995–2006 from 1.2 million tonnes to 2.7 million tonnes[20].

Choosing the best recycling option is not an easy task because each case is different and many different factors have to be taken into account. These include the suitability of material for each waste management option, location, transport, infrastructure, technological developments, economic viability and end markets. It is also important to ensure that the

Key Facts
- Mechanical recycling requires a clean and homogenous waste stream.
- Chemical recycling may be suitable for mixed plastics waste.
- Incineration/combustion of waste plastics recovers their high calorific content if used to generate heat or power.

resources used in the overall recycling operations do not exceed the environmental benefits of recycling. These and other aspects of recycling are discussed in detail later in the book.

1.4.4 Incineration (Without Energy Recovery)

Unlike energy recovery, which reclaims the energy embedded in waste plastics and is hence considered to be a recycling option, incineration without energy recovery only reduces the volume of solid waste and is thus regarded as a waste disposal option. Because it wastes valuable resources, disposal by incineration is considered to be unsustainable. It also raises a number of health and environmental concerns, due to the potential for toxic emissions from combustion (*e.g.* dioxins and heavy metals). However, the latter concerns also exist for incineration with energy recovery and both options are becoming increasingly unpopular with the public.

1.4.5 Landfill

Like incineration, 'landfilling' is also becoming socially unacceptable because of its impacts on the environment and the loss of valuable resources. In addition to these concerns, there is also a problem of finding space for new landfill sites as the existing facilities reach their capacity limits. Hence, waste management policies currently being developed around the world make little allowance for disposal of waste by landfill.

The following sections examine some of these policies and how they affect management of resources.

1.5 RESOURCE AND WASTE MANAGEMENT POLICIES FOR POLYMERS

The key to achieving sustainable resource and waste management involves changing the behaviour of governments, industry and individuals and one way to facilitate change is to design appropriate policies, which maximise resource efficiency and reduce waste generation.

Environmental policies are defined either by legislation or through voluntary agreements between interested parties. Until relatively recently, the emphasis has been on the former and the 'command and control' approach has been predominant. For example, pre 1987 there were 200 command-and-control directives in the European Union (EU). More recently, the emphasis has shifted to the application of economic and market-based instruments such as carbon tax and tradable pollution permits that actively discourage the generation of waste.

Industrial organisations are also instigating various parallel voluntary initiatives ranging from 'waste minimisation', 'zero emission' and 'industrial ecology' projects through 'responsible care' to 'product stewardship' and 'take-back' schemes[9]. They are aimed at improving the environmental performance of industrial activities through the whole life cycle of a product or process. In order to encourage these trends, more progressive governments provide an incentive in the form of financial or other support. This approach, complemented by market-based instruments (*e.g.* carbon tax), provides greater flexibility in the way the targets are achieved and encourages change in industry and society in a more general way than can be achieved by stringent legislation.

However, voluntary agreements are still quite rare and have had only modest success so far[21], thus legislation remains one of the major drivers for more sustainable resource and waste management. We explore both voluntary and legislative aspects of policies and their implementation in different countries in more detail later in the book. Here we continue to give a brief overview of legislation and its implications for polymeric materials and products. As a comprehensive review of legislation around the world is outside the scope of this book, we concentrate below on the EU as an example of how policies have developed and what the future trends in resource management might be in this part of the world.

Key Facts
- Incineration without energy recovery and disposal in landfill wastes natural resources and valuable land space.
- Carbon taxes and tradable pollution permits are designed to discourage the production of waste.

European environmental policy is developed through Action Programmes, which set out action plans related to the environment, usually over a period of 5 years. The Fifth Environmental Action Programme[22] covered the period of 1995–2000 and has now been superseded by the Sixth Action Programme. Both Action Programmes embrace the concept of sustainable development and in particular Agenda 21 (mentioned earlier in the chapter). The Fifth Action Programme adopted the resource and waste management hierarchy shown in Figure 1.4.

Legislation on resource and waste management is one of the key areas of environmental policy development in Europe. It is dominated by the harmonisation of related laws and the development of radical proposals, which encourage more efficient use of resources and re-use of wastes. One of the most important changes in EU policy regarding waste management is the principle of 'producer responsibility'. This policy imposes on producers the obligation to recycle, recover or re-use their products. The development of this policy has been through the imposition of a duty to recover packaging waste and is a flagship for other impeding legislation in this area, including the directives on Waste Electronic and Electrical Equipment (WEEE) and End-of-Life Vehicles, as we shall see below.

The most recent proposal on Integrated Product Policy (IPP) aims to harmonise the existing pieces of legislation and contribute towards more sustainable resource and waste management. IPP would extend the responsibility of manufacturers to cover the environmental impacts of their products throughout their life cycle. It is a natural development from the existing policies on producer responsibility, which are currently concerned with the disposal of waste products at the ends of their lives. The European Commission is currently debating this proposed policy but many questions such as market distortion require resolution before they make firm proposals.

The following sections give an overview of the three EU Directives most directly related to polymer products and materials. The reader interested in further detail on waste legislation in Europe and the UK can consult Appendix 1.

Directive on Packaging and Packaging Waste

This Directive[23] set out to harmonise measures designed to reduce the production of packaging waste, by recovering it in some way, thus reducing the amount remaining for final disposal. Packaging is defined to include products made from any material such as plastic, paper/cardboard, metal, wood and glass, used to contain or protect goods or to assist in their handling, delivery or presentation. The Directive set targets for the recovery of packaging by the year 2001, which included the requirement to recover a minimum of 50 % and a maximum of 65 % of packaging material by weight. Furthermore, it also specified a material recycling rate of 25–45 % (with a minimum allowable figure of 15 % for any single material type) and required the setting up of identification, return, collection and recovery systems.

Directive on End-of-Life Vehicles

The European Parliament and Council Directive of 18 September 2000[24] on end-of-life motor vehicles lays down measures intended to prevent waste from vehicles and provides for the re-use, recycling and other forms of recovery of end-of-life vehicles and their components. Consistent with other European policy, its aim is to reduce the disposal of waste and to improve the environmental performance of all of the economic operators involved in the life cycle of vehicles and especially the operators directly involved in the treatment of end-of-life vehicles. The Directive proposes several recovery, re-use and recycling targets, including recovery and re-use of 85 % by weight of vehicles by 2005, rising to 95 % by 2015. The effect of this Directive will be to force manufacturers to take

> **Key Facts**
> ● Legislation, based on 'Action Programmes' is one of the key areas of environmental policy development in Europe; the concept of 'producer responsibility' makes the producer responsible for waste management.
> ● Directives on packaging, packaging wastes and end-of-life vehicles aim to reduce the disposal of waste and to promote re-use and recycling.

back scrap cars or to meet a substantial part of the cost of recycling. Since plastics are a significant proportion of a car make-up, it also directly affects polymeric materials.

Waste Electrical and Electronic Equipment

Another directive on producer responsibility is the Directive on Waste Electrical and Electronic Equipment (WEEE) adopted by the European Parliament in May 2001[25]. It lays down measures intended to reduce the disposal of waste electrical and electronic equipment through re-use, recycling and other forms of recovery. This will obviously include plastic materials, which make up a large proportion of such equipment. These measures are to be effective within 5 years, with a minimum rate of collection of 4 kg on average *per* inhabitant *per* year by the end of 2005. Its objectives are similar to those contained in the end-of-life directive in that it aims to improve the environmental performance of all economic operators involved in the life cycle of this equipment. It requires producers to provide for the collection of waste electrical and electronic equipment from holders other than private households. When supplying a new product to private households distributors are expected to offer to take back, free of charge, similar waste electrical and electronic equipment in exchange. Suppliers and governments will have to establish systems for the treatment of waste and inspection procedures for the treatment facilities. The proposal also requires the recovery of equipment from private households and other users, and the provision of specific information for both users and for treatment facilities.

The Landfill Directive

The Landfill Directive[26] took 9 years to reach the implementation stage, because of the degree of disagreement and disparity in disposal methods for waste adopted across the European Community. The main thrust of the Directive is the reduction in the amount of biodegradable municipal waste sent to landfill, with the objective of a commensurate reduction in the production of methane gas. The targets for the UK, for example, are reduction by 25 % of the 1995 level by 2010; 50 % by 2013 and 65 % by 2020. The UK has a longer period in which to make the reductions than some other European countries, because of the amount of municipal waste currently being sent to landfill. In the context of polymers, this directive is mainly relevant to the biodegradable plastics, which are currently being developed (see Chapter 8). Added to the fact that biodegradation wastes a valuable resource, this may act as a hindrance to further developments of these types of materials.

In summary, the EU is actively developing resource and waste management policies that have the potential to lead to a more sustainable use of resources. Similar policies are also being developed in other parts of the world, including the USA, Canada and Japan and they will act as a major driver for the recovery and re-use of waste materials. However, their success is also hampered by a number of technical, economic, environmental and social barriers, that limit recovery and recycling of polymeric materials in particular, and which we will discuss further in later chapters. These constraints can only be overcome by a concerted effort from all sections of society, including government, industry and individuals. In the rest of this book we will discuss the role of each group and examine, using practical cases and examples, how polymeric materials and products can be made more sustainable, but first we will explain the structure of the book.

1.6 THE BOOK STRUCTURE AND 'LIFE GUIDE'

We have already highlighted the fact that life cycle thinking is fundamental to sustainable development. We have therefore adopted this approach in analysing the options for re-use and recycling of polymers and it is embedded in the structure of the book. So each chapter follows polymers through different stages in one or more of their life cycles.

Key Facts
● Car manufacturers will be required to take back or pay for disposal of scrap cars by 2015.
● 4 kg of waste plastics in electrical and electronic equipment should be collected *per* person *per* annum in Europe.
● The landfill directive seeks to reduce the amount of biodegradable material sent to landfill and hence reduce methane emissions.

The aim of this chapter has been to provide a 'life guide' by highlighting the issues pertinent to the whole life cycle of polymers in the context of sustainable development. In Chapter 2 we continue on to explore the 'facts of life' and discuss polymer properties and how they may influence their different lives later on. In Chapter 3, we discover how a polymer's 'first life' begins and what happens to polymers when they reach the end of their useful life. 'Second life and beyond' is the subject of Chapter 4, which examines the different recycling options and technologies available for polymers. Chapter 5 discusses 'life forces', or the drivers for recycling and the barriers and how they may be overcome. In Chapter 6, we look at the 'sharp end' of a polymer's life and discuss design for the environment (DFE) as one of the options for reducing the use of resources. In the same chapter we discuss 'life after life' or cascades of uses of polymers, enabled through design for the environment. Then in Chapter 7 we compare the environmental implications of different recycling options and try to identify a 'better life' by comparing the different alternatives. In the eighth and final chapter we look beyond today and wonder what 'life hereafter' might bring for polymers and the implications changing technologies and social patterns could have for the environment and sustainable development.

We hope that you stay with us to discover some (but not all) of the answers to the 'meaning of life' in the context of the impact of polymers on the environment.

1.7 REFERENCES AND FURTHER READING

1. Meadows, D.L. *et al.* (1972). *The Limits to Growth* (Club of Rome), Earth Island, London.
2. IUCN (1980). *World Conservation Strategy*, International Union for Conservation of Nature, Gland, Switzerland.
3. UN Commission (1983). *Common Crisis, North – South: Cooperation for World Recovery*, The Brandt Commission, MIT Press, Cambridge, MA.
4. The Brundtland Commission (1987). *Our Common Future, The Report of the World Commission on Environment and Development (WCOED)*, Oxford University Press, Oxford.
5. UNCED (1992). Agenda 21, United Nations Conference on Environment and Development, June 1992, Rio de Janeiro.
6. CEC (2001). A Sustainable Europe for a Better World: A European Union Strategy on Sustainable Development, COM(2001)264 final. http://europa.eu.int/comm/environment/eussd/index.htm, accessed on 25 July 2001.
7. Moldan, B., Billharz, S. and Matravers, R. (eds) (1997). *Sustainability Indicators: A Report on the Project on Indicators of Sustainable Development*, John Wiley and Sons, Ltd, Chichester, p. 415.
8. McLaren, D. *et al.* (1998). *Tomorrow's World: Britain's share in a Sustainable Future*, Friends of the Earth/Earthscan, London.
9. Azapagic, A. and Perdan, S. (2000). Indicators of sustainable development of industry: a general framework. *Trans. IChemE, B (Proc. Safety Environ. Prot.), Part B*, **78**, (B4), 243 – 261.
10. Azapagic, A. (2001). Life cycle assessment: a tool for choosing sustainable products and processes. In *Green and Sustainable Chemistry*, ed. Clark, J., Macquarrie, D. and Wilson, K., in press.
11. Arentsen, H., Van Lochem and Van Steenderen (1992). In *Polymer Products and Waste Management: A Multidisciplinary Approach*, ed. Smits, M., International Books, Utrecht, p. 28.
12. APME (1999). *Plastics Consumption in Western Europe 1997 – 1999*, APME, Brussels. Also available at: http://www.apme.org/plastics/htm/03.htm.
13. Graedel, T.E. and Alenby, B.R. (1995). *Industrial Ecology*, Prentice Hall, Englewood Cliffs, NJ, p. 412.
14. Jackson, T. (1996). *Material Concerns – Pollution, Profit and Quality of Life*. Routledge, London, p. 218.

15. APME (2001). Association of Plastics Manufacturers Europe, http://www.apme.org/ plastics, accessed on 5 July 2001.
16. Subramanian, P.M. (2000). Plastics recycling and waste management in the US. *Resour. Conserv. Recycling*, **28**, 253–263.
17. ECTEL (1997). *End-of-Life Management of Cellular Phones: An Industry Perspective and Response*, Report of the ECTEL Cellular Phones Takeback Working Group, November 1997.
18. Clift, R. and Wright, L. (2000). Relationships between environmental impacts and added value along the supply chain, *Technol. Forecasting Social Change*, **65**, 281–295.
19. Clift, R. and Longley, A.J. (1994). Introduction to clean technology. In *Clean Technology and the Environment*, ed. Kirkwood, R.C. and Longley, A.J., Blackie Academic and Professional, Glasgow, Chapter 6.
20. APME (1998). Options for plastics waste recovery, http://www.apme.org/environment/ htm/06.htm, 10 December 1998; accessed on 5 July 2001.
21. Nunan, R. (1999). Barriers to the use of voluntary agreements: a case study of the development of packaging waste regulation in the UK, *Eur. Environ.*, **9**, 238–248.
22. EC (1992). Resolution of the Council on the Fifth Community Policy and Action Programme on the Environment and Sustainable Development (1993–2000), *Offic. J. Eur. Commun.*, **C138**, 17 May 1993.
23. EC (1994). Council Directive 94/62/EC of 20 December 1994 on Packaging and Packaging Waste, *Offic. J. Eur. Commun.*, **L 365**, 10–23.
24. EC (2000). End of life Vehicle Directive, http://ue.eu.int/newsroom/ Document No. 8828/ 00 (Presse 179), dated 23 May 2000, accessed 13 November 2000.
25. EC (2000). The Waste Electrical and Electronic Equipment directive, http://europa. eu.int/eur-lex/en/com/dat/2000/en_500PC0347_01.html (2001).
26. EC (1999). Council Directive on Landfill of Waste (99/31/EEC). *Offic. J. Eur. Commun.*, **L 182**, 16 July 1999.

Further Reading

Auty, R.M. and Brown, K. (eds) (1997). *Approaches to Sustainable Development*, Pinter, London.
Carley, M. and Christie, I. (2000). *Managing Sustainable Development*, 2nd edn, Earthscan, London.
Jackson, T. (1996). *Material Concerns – Pollution, Profit and Quality of Life*, Routledge, London, p. 218.
Kirkwood, R.C. and Long, A.J. (eds) (1995). *Clean Technology and the Environment*. Blackie Academic & Professional, London.
Lafferty, W.M. and Eckerberg, K. (eds) (1998). *From the Earth Summit to Local Agenda 21: Working Towards Sustainable Development*. Earthscan, London.
Maslow, A. (1970). Motivation and Personality, 2nd edn, Harper and Row, New York.
Pickering, K.T. and Owen, L.A. (1994). *An Introduction to Global Environmental Issues*. Routledge, London, p. 390.
Scott, G. (1999). *Polymers and the Environment*. Royal Society of Chemistry, Cambridge.
Smits, M. (ed.) (1996). *Polymer Products and Waste Management – A Multidisciplinary Approach*. Internation Books, Utrecht, p. 256.
The Engineering Council (1994). *Guidelines on Environmental Issues*, The Engineering Council, London, p. 56.
Yakowitz, M. (1997). *Sustainable Development: OECD Approaches for the 21st Century*, OECD, Paris.

1.8 REVISION EXERCISES

1. Define sustainable development in your own words and list five global economic, social and environmental issues that need to be addressed urgently. Explain how you think they could be solved.

2. One of the objectives of sustainable development is the satisfaction of human needs. Make a list of the needs that you personally would like to satisfy. Now compare this with Maslow's heirarchy of needs (see Further Reading). Compare your priorities with your friends and discuss the differences. On a global level, how do you think these priorities differ between different countries and cultures? What does that tell you about how easy or difficult it is going to be to satisfy everyone's needs? And how about future generations?

3. Explain what you understand by 'life cycle thinking'. Why is that important for sustainable development?

4. What is Life Cycle Assessment? How is that different from 'life cycle thinking'?

5. Visit the APME web site and answer the following question: How can plastic materials contribute to sustainable development? Give examples of how plastics contribute to the environmental, economic and social components of sustainable development.

6. If plastic materials contribute to sustainable development, why are they an issue?

7. Summarise the options in the resource and waste management hierarchy and give examples relevant to polymeric products and materials for each option.

8. How can government, industry and individuals help towards more sustainable use of resources? Support your answers by giving examples relevant to polymeric materials.

9. Which EU Directives are directly related to polymeric materials? How do you think they are going to affect the use of polymers in the future?

polymers in everyday use: principles, properties and environmental effects

Chapter 2 – The Doom Fulfilled (Perseus Slaying the Sea Serpent) (E Burne-Jones, 1875 – 77).
After vanquishing the Medusa and escaping with her severed head, Perseus journeys to Joppa, where he finds Andromeda, daughter of Cassiopeia, chained to a rock as a sacrifice to a sea monster. In the best heroic tradition, Perseus slays the sea monster, rescues Andromeda and later marries her. The link between the long, coiling polymer (here represented metaphorically by the sea serpent) and the environmental agents of its destruction (symbolised by Perseus) is more tenuous, but this painting had to be included as it's one of the editor's favourites!

2.1 INTRODUCTION

In this, the 21st century, polymers are ubiquitous in our everyday environment. As illustrated in Table 2.1, they come in myriad forms, with a wide variety of chemical structures and a bewildering array of properties, which can often be obtained by relatively subtle changes in preparative or processing chemistry. This allows the polymer scientist the opportunity to tailor both structure and properties almost at will and makes the polymer a most versatile functional material. The resulting product may be both lightweight and also of high specific strength when compared with conventional structural materials such as wood, metals or glass. At the same time, greater processability and greater durability or longevity in highly aggressive environments mean that the polymer is economically attractive to fabricate and use. However, the greater durability may prove a double-edged sword when it becomes necessary to recycle the polymer (after the primary or secondary application) or dispose of it (at the end of its useful lifetime).

Key Facts
- Polymer properties can be tailored to purpose by the processing chemistry.

Table 2.1 Different categories of polymers and their physical characteristics.

Polymer category[a]	General characteristics	Typical example	Typical uses
Natural elastomers	Readily undergo deformation and exhibit large, reversible elongations under small, applied stresses (elasticity)	Poly(*cis*-isoprene) 'natural rubber'	General purposes, car tyres
Synthetic elastomers		Poly(acrylonitrile-*co*-buta-diene-*co*-styrene) (ABS)	Gaskets, flexible fuel tanks and oil hoses
Natural fibres	Resistant to deformations and characterised by a high modulus and low percentage elongations	Cellulose	Paper, textiles (as cotton)
Synthetic fibres		Poly(hexamethylene adipamide), *e.g.* nylon 6,6	Textiles, carpets, tyre cord
Commodity thermoplastics	Capable of changing shape on application of force and retaining this shape on removal of force (stress produces a nonreversible strain). Will soften when heated above T_g and can be reshaped and will harden in this form upon cooling	Polystyrene	Wall-tiles, flowerpots, beverage cups
Engineering thermoplastics		Polyimides	Microelectronics, structural composites
Commodity thermosets	Become permanently hard when heated above critical (cure) temperature and will not soften again on reheating. Insoluble once in this cross-linked state	Amino resins (melamine – formaldehyde)	Coatings, laminated surfaces
Performance thermosets		Epoxy resins	Adhesives, structural composites

[a]It is important to note that there is often no clear distinction between the categories listed in Table 2.1 and that some polymers may belong to more than one classification. For instance, polypropylene (a typical thermoplastic) may also form fibres, while polyurethanes may be elastomers or plastics depending on their molecular structure. Furthermore, while the term 'plastic' is often used to describe polymers, the words 'polymer' and 'plastic' are not synonymous and the latter should only be used to describe the stress–strain behaviour rather than refer to the chemistry of the materials.

In this chapter we will concentrate primarily on the fundamental aspects of polymer properties and chemistry (*e.g.* synthesis, structure and degradation) that are relevant to polymer re-use and recycling. Owing to the nature of the commodity polymer market, where five thermoplastic polymers account for around 75 % of the total polymer consumption (thermoplastics and thermosets combined), most of the chapter, and indeed a significant portion of the book, will address the issues surrounding the recycling of thermoplastic polymers. This is not to lose sight of (nor underplay) the vital technological role that thermoset polymers play in today's society. In some cases even engineering thermoplastics (whose thermal and mechanical performance is superior to commodity thermoplastics) are unable to compete with the balance of properties offered by higher performance thermoset polymers. However, the greatest demand will continue to be focused on the need to manage waste streams containing large tonnages of thermoplastic polymers. The main characteristics of and differences between thermoplastics and thermosets are highlighted in Textbox 2.1.

Textbox 2.1 Thermoplastics and thermosets

The majority of commercial polymers in common use are *thermoplastics*: polymers that can generally be safely processed several times by melting and shaping the melt and the final product is obtained by cooling. These are generally either linear or branched polymers so that there are few chemical interactions between chains. In contrast, thermosetting polymers (*thermosets*) will undergo irreversible reaction (*cure*) on heating and this is usually accompanied by the formation of covalent bonds between polymer chains (*cross-links*). The resulting three-dimensional network loses its solubility and does not exhibit a melting point, since the individual chains may no longer flow past one another. Thermosets may soften when heated, *e.g.* when the glass transition temperature (T_g) of the polymer is reached and the glassy polymer becomes more rubber-like, although in highly

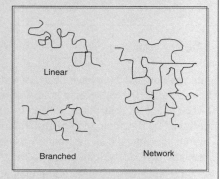

Figure 2.1 Common polymer structures

cross-linked polymers the T_g may be so high that thermal degradation (the breaking of covalent bonds) occurs before this event. The T_g is clearly an important parameter, especially in the case of amorphous polymers, and quite sensitive to changes in structure. We will discuss the factors influencing the magnitude of T_g later in this chapter. A range of polymer architectures is possible and some common forms are shown in Figure 2.1.

Key Facts
● Thermoplastic polymers can be processed and re-processed several times.

As the terminology used in this chapter on occasion presupposes some prior knowledge, a list of selected reference titles is presented at the end of the chapter to aid your understanding. These textbooks may provide either grounding in polymer science or simply a revision of previous study.

2.2 BASIC PRINCIPLES OF POLYMER CHEMISTRY

2.2.1 Nomenclature and Basic Definitions

In essence, polymers are large molecules made up of relatively simple repeating units ('building blocks'), known as monomers. The term 'polymer', first coined by the Swedish chemist Jöns Jakob Berzelius, in 1833, is derived from the Greek *poly* (many) and *meros* (part), while *monomer* describes a single repeat unit. In the chemical literature, the terms polymer and *macromolecule* (literally 'large molecule' from *makromolekül* used by the German chemist Hermann Staudinger in 1924) are often used interchangeably. However, while a polymer is by definition a macromolecule, the latter may not necessarily be a polymer (*i.e.* made up of repeating monomer units). It follows that the process by which monomers (having at least two reactive functional groups) undergo reaction to produce polymers is known generically as polymerisation. The resulting polymer may be represented by a *constitutional repeat unit* or CRU (the smallest structural motif that may be repeated to make up the polymer chain). In the case of an addition polymer, this may simply be the monomer unit enclosed by square brackets or parentheses (Figure 2.2), although in the case of polyethylene the CRU is $-CH_2-$ (methylene) rather than $-CH_2CH_2-$. A subscripted number, *n*, is usually placed after the second bracket to indicate the number of repeat

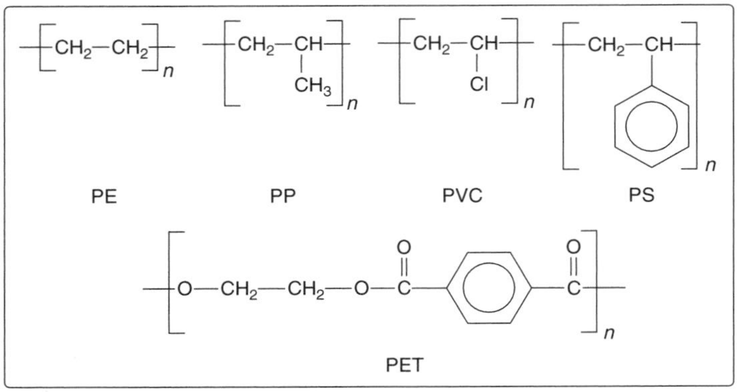

Figure 2.2 Structures of some common commercial polymers showing constitutional repeat units

> **Key Facts**
> ● DP (average degree of polymerisation) =
> $$\frac{\sum (\text{number of chains} \times \text{length})}{\sum \text{number of chains}}.$$

units (monomers) that have been used in the construction of the polymer chain (*i.e.* the chain length). The value of *n* is also termed the *average degree of polymerisation*, or $\overline{DP}$. The bar indicates that this is a mean value calculated as the (sum of the number of chains multiplied by the chain length)/(total number of chains).

While polymers may be named using the International Union of Pure and Applied Chemistry (IUPAC) system, based on the monomer structure, the nomenclature of polymers is not fully systematic and, in practice, a mixture of common (trivial) names, trade names and acronyms is used. Usually the monomer is simply prefixed with *poly* (or enclosed in brackets if it is made up of more than one word). For example, PET is commonly called poly(ethylene terephthalate), but more correctly should be systematically named poly(oxy ethylene oxy terephthaloyl). In this chapter we use both the acronyms and the names with the prefix *poly*. In the rest of the book, for simplicity acronyms are used more predominantly (the acronyms are explained in Appendix 4 at the end of the book).

As the number of monomers increases in the chain, the terms '*dimer*' (containing two monomers), '*trimer*' (three monomers), and '*tetramer*' (four monomers), *etc.* are used. The term '*oligomer*' is used rather loosely to define a small polymer containing several units (*oligos* = few). Bulk polymers are composed of polymer chains (macromolecules) containing several thousands to many hundreds of thousands of monomers. It is the variety of properties, and particularly the macroscopic properties, which these polymers exhibit (*e.g.* their viscous-flow behaviour called *viscoelasticity*) that sets them apart from their low molecular weight analogues.

As we have already noted, the number of polymers available is enormous and their discussion is outside the scope of this book. However, for your reference, Table 2.1 lists the main types of polymer and their major uses. Different types of common thermoplastic materials considered in the book are also listed in the table.

2.2.2 Polymerisation Methods

Historically, the polymer industry has relied heavily on nonrenewable, fossil resources for the necessary feedstock materials Figure 2.3. In the first instance, this was coal (which yielded tars and acetylene) and then oil and petroleum (which, after processing and thermal 'cracking', produces ethylene and other monomers). Almost all polymers in everyday use such as plastics, rubbers and fibres are synthesised from chemicals derived from oil (and the price of this commodity has a significant impact on the economics of both plastics manufacture and recycling). The chief exceptions to this rule are natural rubber and related polymers derived from *caoutchouc*, and polymers derived from cellulose, such as cellulose acetate.

Monomers are then used to synthesise polymers, perhaps in the presence of a catalyst (*e.g.* benzoyl peroxide or azoisobutyronitrile in the case of free radical polymerisation of a 1-alkene) and some form of energy source: typically heat (although more efficient sources

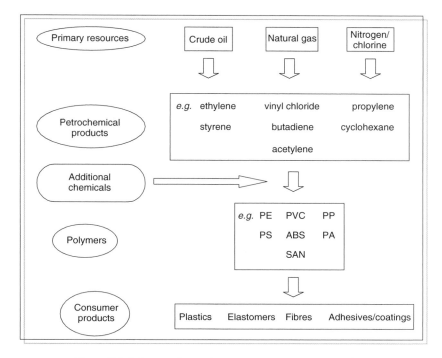

Figure 2.3 Polymers: from primary sources to consumer products

are being explored and are discussed in Chapter 8). The monomer is processed either in the bulk as a molten material, or as a solution or an emulsion and these different routes are discussed in the context of common commodity polymers in Chapter 3. In the molten state, the viscosity or melt flow index are important when perhaps introducing monomers into a mould or incorporating reinforcing fibres in the case of a composite. There are two generic polymerisation processes, each occurring with many variations: step growth and chain growth polymerisation.

(i) Step growth polymerisation can occur *via* a series of condensation reactions (hence the older name *condensation polymerisation*), although this is not always the case. However, in most cases, the resulting polymer differs slightly from the original monomer(s) as a small molecule is eliminated during the reaction. The formation of PET occurs *via* a step growth mechanism (to liberate water) and this is shown in Figure 2.4.

(ii) Chain growth polymerisations tend to occur through the reaction of multiple bonds initiated by a free radical or ion. The product of a chain growth (or *addition*)

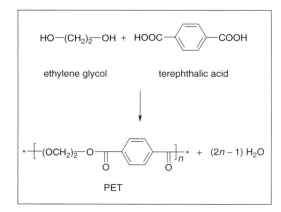

Figure 2.4 Preparation of poly(ethylene terephthalate) (PET) by step growth polymerisation

Figure 2.5 Preparation of polypropylene (PP) by chain growth polymerisation

Key Facts
● Polymers are synthesised by step growth (condensation) or chain growth (addition) polymerisation.
● Many polymers have a structure consisting of amorphous and crystalline domains.

polymerisation has the same chemical composition as the monomer. For example, a simplified polymerisation scheme for polypropylene (PP) is given in Figure 2.5 and no small molecules are eliminated during the process. The (*) symbol indicates that the chain continues beyond the constitutional repeat unit. The PP polymer is shown in a 'head-to-tail' configuration and this phenomenon (orientation) is explained later in the Section 2.2.4.

These processes have been presented in a drastically simplified manner for the purposes of this chapter and a full discussion of the polymerisation processes is outside the scope of this book (you may wish to examine the bibliography for further information). However, the basic differences inherent in these polymerisation mechanisms are outlined in Table 2.2.

Table 2.2 Comparison of the general characteristics of step growth and chain reaction polymerisation methods.

Chain reaction	Step reaction
Growth occurs by successive addition of monomers to a limited number of growing chains	Growth occurs throughout, from reaction between monomers, oligomers, and polymers
Monomer is consumed relatively slowly, but molecular weight increases rapidly	Monomer is consumed rapidly, but molecular weight increases slowly
Ultimate degree of polymerisation can be very high	Ultimate degree of polymerisation is either low or moderate
Reaction comprises discrete initiation and propagation mechanisms	No initiator required; same reaction mechanism occurs throughout reaction
Chain termination usually occurs	No termination step; end groups remain active
Polymerisation rate increases initially as initiator is activated; remains relatively constant until monomer is consumed	Polymerisation rate decreases steadily as functional groups are consumed

2.2.3 Polymer Morphology

The subject of polymer morphology (in essence the physical and mechanical properties displayed by the polymer) is of great importance for the study of polymers. It may influence the solubility of the polymer (and hence have an impact on the processing of the material), the softening temperature (and may therefore govern the maximum use temperature) and even the gas permeability (perhaps making a polymer unsuitable for some packaging applications, where carbonated beverages are concerned). In this section we will discuss crystallisation (and crystallinity), the glass transition temperature and the factors that affect its magnitude. Not all polymers exhibit a melting transition when heated (see Textbox 2.1), but some of those that do may crystallise as they are precipitated from solution or from the molten state (although most polymers tend not to crystallise on cooling but rather to vitrify to form a glassy solid). We shall discuss crystallinity later in this chapter, but the greater the degree of attraction between discrete polymer chains, the greater is the degree of crystallinity. However, the tendency to crystallise is opposed by the irregularity of the polymer structure, particularly when considering highly branched polymers such as low density polyethylene, LDPE. The branching disrupts the packing forces, such as dispersion forces (and dipole–dipole interactions, or hydrogen bonds in polymers containing heteroatoms), which cause the polymer chains to associate and enhance crystallinity. The growing molecular weight of the polymer increases the likelihood of chain entanglement and raises its viscosity, which, in turn, reduces the rate of diffusion. This may lead to a reduction in the rate of polymerisation as reactive

A modern Polymerisation Plant

monomers and oligomers are prevented from meeting to undergo reaction. Consequently, unless a polymer is highly *amorphous* (*i.e.* lacking long-range order) or *glassy*, then typically it will contain both ordered crystalline domains and disordered or amorphous domains, which may lead to the polymer exhibiting more than one glass transition temperature or a broad melting temperature. Not only does the fundamental chemical structure affect the balance of the domains (the *degree of crystallinity*), but it can also influence the way in which the polymer is processed. It may be useful to produce polymers that exhibit a high melting temperature and perhaps a large difference in temperature between the glass transition temperature, T_g, and the melting temperature, T_m. For instance, this is particularly important in the case of polyamides (nylons) where it is important that the polymer textile fibres undergo softening during ironing without undergoing melting. For maximum crystallinity, polymers should have molecular structures that are able to pack together easily. We will examine this further below when discussing polymer stereochemistry. While some polymers may experience a true change of phase during melting (T_m), the more common glass transition (T_g), simplistically representing the change from a rigid plastic below T_g to a flexible, rubbery material above T_g, is not so well understood. As the polymer is solid both before and after the transition has occurred, T_g is termed a second-order phase change, although it has a profound effect on the nature and magnitude of mechanical and physical properties. A general rule is that the ratio of T_g/T_m lies in the range 0.5–0.8.

When considering a simple polymer (in the absence of a plasticiser), T_g represents the temperature range in which the polymer is experiencing rotation of segments of the chains at the molecular level and hence undergoing softening at the macroscopic scale. For this to happen, the polymer must have sufficient thermal energy to overcome the activation energy for rotation and sufficient free volume (*i.e.* the voids between chains) to allow the rotation of segments to take place without chain entanglement occurring. The magnitude of T_g often governs the maximum temperature at which a polymer may be used (as they are often used as structural materials where mechanical properties may be of paramount importance). For instance, expanded polystyrene (EPS) is used to fabricate drinking cups. The T_g of EPS is just below 100 °C, making it suitable for hot (but not boiling) beverages.

As the phenomenon of chain rotation involves molecular motion, the magnitude of T_g is influenced by a number of molecular/structural features that either increase the association between chains and/or reduce the mobility/free rotation of individual segments of the chains. The features that affect T_g are listed and discussed in Textboxes 2.2a–d. Note that the value of T_g is not absolute and usually covers a range of temperatures. It is also a function of molecular weight; consequently, the representative values presented below are merely a guide for the trends experienced. It is also apparent that some of these features may also influence the degree of crystallinity that a polymer exhibits.

Key Facts
● Nylon must be softened during ironing but not so much that it melts so it must have a low glass transition temperature (T_g) and a high melting temperature (T_m), *i.e.* towards the top of the typical T_g/T_m range of 0.5–0.8.
● The T_g of expanded polystyrene is just below 100 °C, making it unsuitable for drinking cups for boiling beverages.
● T_g is influenced by the activation energy of rotation of monomer units and the amount of free space available in the polymer.

Textbox 2.2a Chain flexibility

Although it may be difficult to examine the effect of this factor on T_g in isolation (aside from the effects of side group substitution and steric hindrance, discussed below), it is probably the single most important factor involved in determining the magnitude of T_g. The lower the barrier to rotation the lower the value of T_g: this may be illustrated by looking at homochain polymers (containing carbon–carbon backbones) and heterochain polymers, containing heteroatoms in the backbone (Figure 2.6). This in turn may make the process of recycling polymers simpler and easier

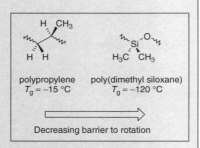

polypropylene
$T_g = -15$ °C

poly(dimethyl siloxane)
$T_g = -120$ °C

Decreasing barrier to rotation

Figure 2.6 Influence of chain flexibility on the glass transition temperature T_g (at atmospheric pressure)

to achieve when a thermal process is involved. If the overlap of the atomic orbitals on adjacent atoms in the backbone is significantly reduced, then the barrier to rotation is reduced. The incorporation of, for instance, an aromatic ring into the polymer backbone stiffens the chain and raises T_g.

Textbox 2.2b Side group effects

A combination of size and bulkiness (and hence steric hindrance) and rigidity/flexibility is important for T_g. As the steric hindrance is increased by the substitution of increasingly bulky side groups, then the T_g is raised (Figure 2.7). If the side chain becomes increasingly flexible then this makes the polymer more flexible and initially leads to a reduction in T_g as the chains are plasticised and the interaction between chains reduced (Figure 2.8). However, as the chain length increases still further, the T_g may rise. This observation seems counterintuitive, but as the longer side chains may become more entangled then they serve to increase the association between the chains and T_g rises once more.

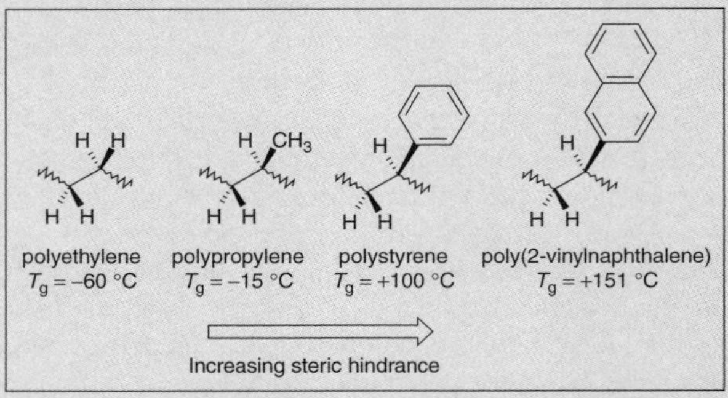

polyethylene $T_g = -60\ °C$ polypropylene $T_g = -15\ °C$ polystyrene $T_g = +100\ °C$ poly(2-vinylnaphthalene) $T_g = +151\ °C$

Increasing steric hindrance

Figure 2.7 Influence of steric hindrance on the glass transition temperature T_g

$T_g = 110\ °C$ $T_g = 65\ °C$ $T_g = 35\ °C$ $T_g = 20\ °C$

Increasing side group flexibility

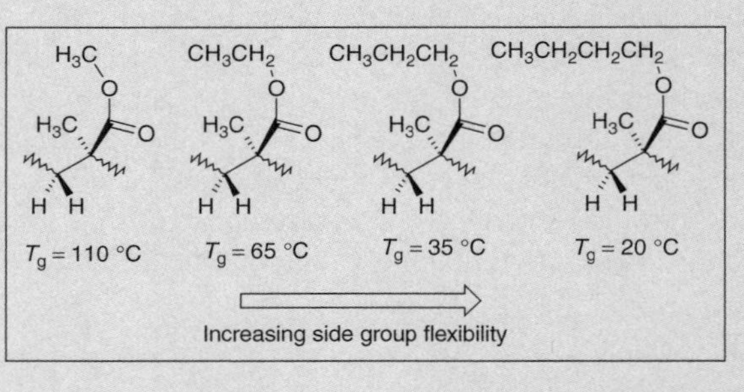

Figure 2.8 Influence of side group flexibility on the glass transition temperature T_g

Textbox 2.2c Symmetry and polarity

Symmetry: an increase in molecular symmetry tends to reduce T_g. This is probably related to the amount of free volume generated. As the chains can pack less easily, more free volume is generated and less thermal energy is required to bring about chain rotation (Figure 2.9).

Polarity: an increase in polarity tends to raise the T_g value, as a result of the increase in dipole–dipole association that occurs between neighbouring chains (Figure 2.10).

Hydrogen bonding is perhaps a more specific example of increasing the magnitude of the inter-chain associations. Figure 2.11 shows the effect of substituting an amido group

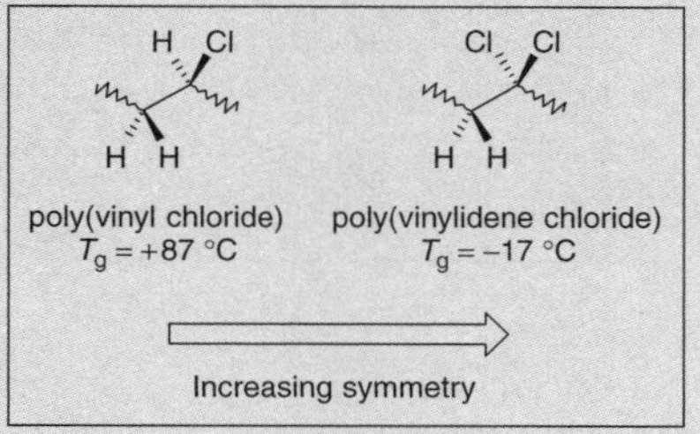

Figure 2.9 Influence of molecular symmetry on the glass transition temperature T_g

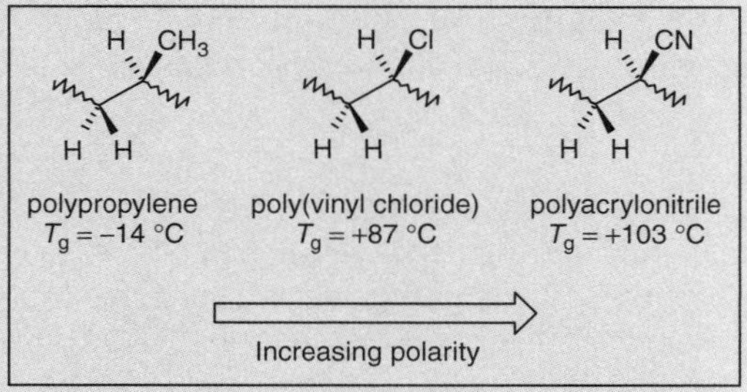

Figure 2.10 Influence of chain polarity on the glass transition temperature T_g

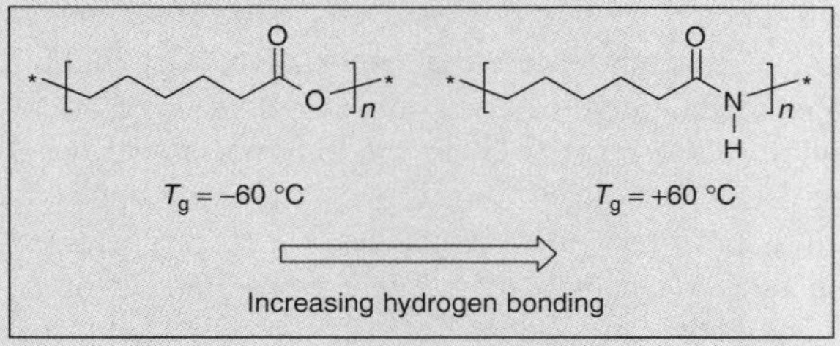

Figure 2.11 Influence of hydrogen bonding on the glass transition temperature T_g

(which can form hydrogen bonds between the nitrogen and hydrogen atoms on adjacent chains) into the main chain for an ester link (which is unable to form a hydrogen bond in this instance).

Key Facts
- Reducing overlap of atomic orbitals in between adjacent atoms in the backbone reduces the barrier to rotation and reduces T_g.
- Adding bulky side groups increases T_g.
- Increasing symmetry reduces T_g.
- Increasing polarity increases T_g.
- The T_g of a copolymer is normally between the T_gs of the homopolymers.
- Cross-linking increases T_g.

Textbox 2.2d Copolymerisation and cross-link density

Copolymerisation: generally the T_g of a copolymer lies somewhere between the T_g values of the pure homopolymers. The introduction of a comonomer tends to disrupt the packing of the pure homopolymer, leading to a reduction in the T_g of the higher component. The composition of the copolymer is important in determining the resulting value for the copolymer (T_{gc}), since:

$$1/T_{gc} = W_1/T_{g1} + W_2/T_{g2} \qquad (2.1)$$

$$T_{gc} \text{ (Kelvin)} = V_1 T_{g1} + V_2 T_{g2} \qquad (2.2)$$

where W_1 and W_2 are the weight fractions of each comonomer, V_1 and V_2 are the corresponding volume fractions, T_{g1} and T_{g2} are the glass transition temperatures in Kelvin of the individual homopolymers. Equation (2.1) tends to give a curved plot that slightly underestimates the value of T_{gc}, whereas Equation (2.2) (yielding a straight line) tends to overestimate the value and so the true T_{gc} tends to lie between the two. This allows the value of T_{gc} to be tailored conveniently by varying the composition.

Cross-link density: as cross-linking is increased the mobility of the chains becomes greatly reduced as covalent bonds are formed between them and T_g is increased.

Key Facts

● Bulk properties of polymers are directly affected by stereochemistry.
● Elasticity depends on 'light' cross-linking of polymer chains.
● Monomer units in a polymer chain can orient head-to-head or head-to tail.

2.2.4 Polymer Stereochemistry

The topic of *stereochemistry* (the three-dimensional spatial arrangement of atoms in molecules) has great importance for conventional organic compounds. In a polymer, the differences in microstructure take on an even greater significance, because the polymer's bulk properties, such as melting temperature or solubility in organic solvents, are directly affected by its stereochemistry. Several kinds of variation can be identified and these are discussed below.

Polymer Architecture

A variety of polymer architectures is possible, depending on the way in which the covalent bonds are formed between the atoms in the chains. Linear and branched forms are most commonly found in commercial thermoplastics (see Textbox 2.1). Thermosetting polymers tend to form amorphous, three-dimensional networks in which chemical 'cross-links' are formed between polymer chains, anchoring them together. Elastomers, *i.e.* polymers like natural rubber, poly(*cis*-1,4-isoprene), display intermediate forms in which linear or branched chains are lightly cross-linked to tether the chains together, while allowing some translational movement. It is this feature that conveys *elasticity* to the polymer: the chains can slip past one another, but are restrained from complete dissociation (and the formation of a molten polymer) by the cross-links. In the bulk polymer this behaviour allows the material to undergo some significant deformation under the application of a deforming stress, but to return to its original dimensions when the force is released (provided that the yield point is not exceeded and that the covalent cross-links remain intact). Other, more exotic, architectures are also possible, but are rare in most commercial polymers. These forms are quite descriptive and include:

● *star* polymers (chains grow outwards from a single, central core, such as dendritic polymers or 'dendrimers');
● *ladder* polymers (backbones are linked by covalent 'rungs');
● *comb* polymers (a number of side chains are grafted on to a single backbone to resemble a hair comb).

Polymer Orientation

In one of the examples given previously, the PP chain was shown in a 'head-to-tail' orientation. During a free radical polymerisation (in which the initiation and propagation steps involve radical species) the radical (R$^\bullet$) may attack an asymmetric vinyl monomer in

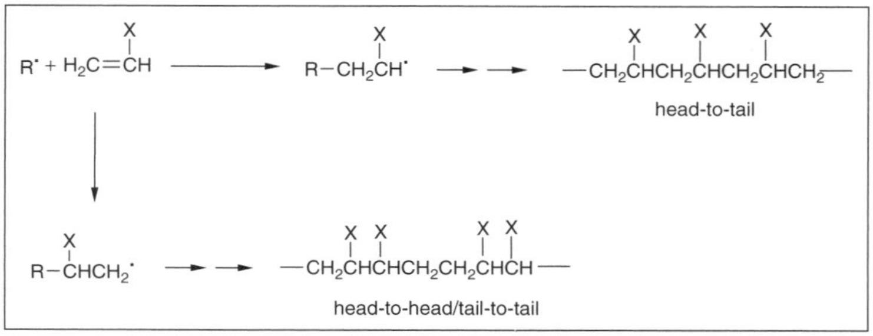

Figure 2.12 Free radical polymerisation of vinyl monomer (bearing substituent X) showing different possible orientations in the polymer

one of two ways, leading to two possible configurations of the monomer unit (Figure 2.12). If the radical attacks the carbon bearing the substituent X, then a 'head-to-head' chain ensues; if the other carbon atom is attacked then a 'head-to-tail' chain is formed. The position of attack depends largely on the stability of the product formed (*e.g.* whether substituent X can take part in resonance stabilisation with the unpaired electron on the adjacent α-carbon atom) and the steric hindrance caused by the substituent X (to repel the attack of the radical). In the majority of polymers, head-to-tail chains are formed, but there are some exceptions, *e.g.* poly(vinylidene fluoride-*co*-vinyl fluoride), which may contain up to 32 % head-to-head links.

Polymer Configuration

It is possible to effect very significant changes to a polymer's properties by quite subtle alterations in the way in which the atoms and monomers are configured (*i.e.* how they are bonded) in the polymer chain. For instance, the T_g exhibited by polypropylene may be significantly altered from $-4\,°C$ (syndiotactic PP), to $-6\,°C$ (atactic PP) or $-18\,°C$ (isotactic PP) by changing the relative positions of the methyl (CH_3) substituent along the polymer backbone. Other polymers may show even greater changes (*e.g.* syndiotactic and atactic PMMA display a T_g of $105\,°C$, while the isotactic form of PMMA has a T_g of $38\,°C$).

Figures 2.13–2.15 show the polymer arising from the substituted vinyl alkene ($CH_2{=}CHCl$) to illustrate polymer configuration. You may remember that each carbon centre in the saturated polymer backbone is sp^3 hybridised and therefore the substituents are arranged tetrahedrally. This means that the growing chain can adopt three different possible spatial arrangements (*atactic, isotactic* and *syndiotactic*).

> **Key Facts**
> ● The distribution of substituents of the chain may be random (atactic), on alternate sides (syndiotactic) or along one side (iostactic), and has a major influence on the polymer bulk properties.

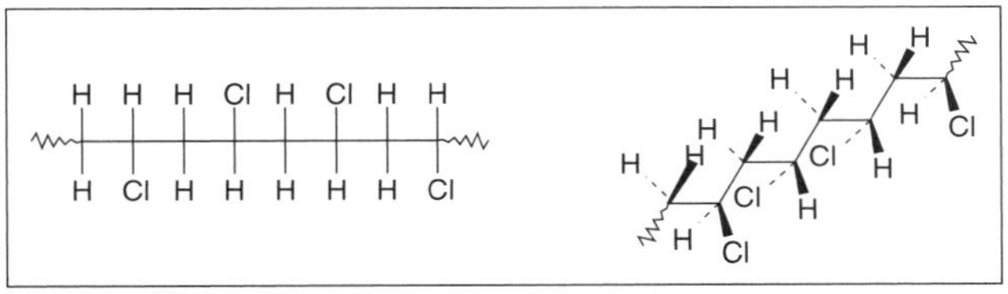

Figure 2.13 Atactic spatial arrangement (substituents arranged randomly along chain)

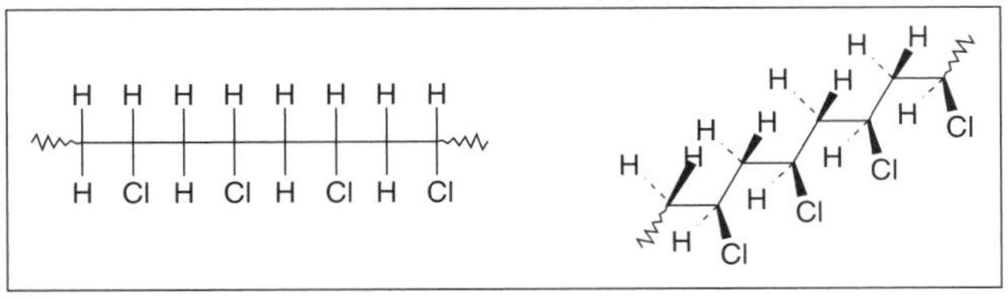

Figure 2.14 Isotactic spatial arrangement (substituents arranged along one side of chain)

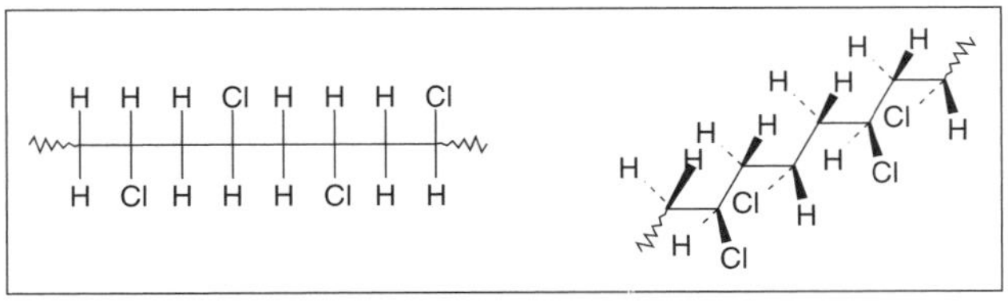

Figure 2.15 Syndiotactic spatial arrangement (substituents arranged on alternate sides of chain)

Iso- and syndiotactic polymers are often highly crystalline with superior mechanical properties and these materials are now generally prepared using stereospecific catalysts (such as Ziegler–Natta catalysts). These catalytic species act by controlling the mode of entry of the monomer unit to a growing chain using steric hindrance, electrostatic interaction or coordination to a metal atom.

2.2.5 Average Molar Masses and Distributions

A knowledge of the average molar mass enables a determination of the polydispersity index (or molar distribution) to be made and these topics are discussed in the subsequent section. The latter is of particular interest when examining the properties of polymers and their processability. It is usually impractical (without resorting to time-consuming fractionation techniques) to synthesise polymers of a single molar mass (chain length) and so almost all synthetic polymers, and some naturally occurring polymers, comprise a mixture of individual polymer molecules of a variety of chain lengths. For such mixtures, it becomes inappropriate, although not inaccurate, to talk of relative molar mass (RMM). Instead, a statistical approach is taken to define the average molar mass and the breadth of this molar mass distribution (the *polydispersity index*). Historically, the analytical techniques that have been used to measure molar masses of polymers fall into two categories:

(a) techniques which count particles, perhaps based on colligative properties (*e.g.* viscosity in solution), leading to the number average ($\overline{M}_n$) molar mass (Equation 2.3);
(b) techniques which measure the size of particles, such as light scattering, leading to the weight average ($\overline{M}_w$) molar mass (Equation 2.4). These two averages have persisted and are still commonly used to define molar masses in polymer science.

Key Facts
- Number average molar masses are based on colligative properties such as viscosity.
- Weight average molar masses are based on particle size and measurement techniques such as light-scattering.
- $\overline{M}_w$ is normally higher than $\overline{M}_n$, because larger chains have a greater mass than smaller chains.
- The viscosity average molar mass is dependent on the hydrodynamic volume of the polymer in solution.

Number Average Molar Mass ($\overline{M}_n$)

The number average represents the common arithmetic mean and is written in terms of the *number* of molecules of a particular RMM (the number average molar mass, $\overline{M}_n$):

$$\overline{M}_n = \sum N_i M_i / \sum N_i \tag{2.3}$$

where N_i is the number of chains (molecules) with an RMM of M_i and $\sum N_i$ is the total number of chains.

Weight Average Molar Mass ($\overline{M}_w$)

The weight average molecular weight, on the other hand, is obtained by recording the total mass of each chain of a particular length, summing these masses, and dividing by the total mass of the sample.

$$\overline{M}_w = \sum w_i M_i / \sum w_i = \sum N_i M_i^2 / \sum N_i M_i \tag{2.4}$$

Here $w_i = N_i M_i$ and is the mass of chains with an RMM of M_i.

When determining $\overline{M}_w$, the larger chains in a sample have greater mass than the smaller chains and consequently skew the weight average to higher values. As a result $\overline{M}_w$ is always greater than $\overline{M}_n$, unless the polymer sample is *monodisperse* (*i.e.* all the chains are of an identical length) which, as we have said, is almost never the case.

Viscosity Average Molar Mass ($\overline{M}_v$)

There is a third, commonly used, average called the viscosity average molar mass. This is obtained from viscosity measurements of dilute solutions made in a glass U-tube viscometer. The basis of the technique is that there is a relationship between the relative magnitude of the increase in viscosity and the molar mass of the polymer. Relatively small changes in the size or conformation of a polymer chain will affect its frictional properties (when dissolved in a liquid, *i.e.* its *hydrodynamic volume*) and this will alter the measured time taken for a polymer to flow through a glass capillary tube. The viscosity average molar mass can be derived from the Mark–Houwink equation:

$$[\eta] = K_v \overline{M}_v \tag{2.5}$$

where $[\eta]$ = limiting viscosity number (characteristic for the polymer in a particular solvent) or 'intrinsic viscosity at zero concentration', K_v and v are also characteristic for a polymer/solvent system and can be obtained by calibrating with fractions of known molar mass. This then enables the molar mass of an unknown polymer fraction to be obtained simply by determining (experimentally) $[\eta]$ for several different polymer concentrations and plotting log $[\eta]$ against log (molar mass); interpolation then gives the molar mass that corresponds to the measured intrinsic viscosity.

The viscosity average molar mass is defined as:

$$\overline{M}_v = \left[\sum N_i(M_i^{1+v}) / \sum N_i M_i\right]^{1/v} \tag{2.6}$$

where the viscosity exponent v is dependent on the nature of the polymer, the solvent system and the analysis temperature. The value of v is a measure of the ease with which the polymer is dissolved. Typically, for a polymer dissolved in a *theta* solvent (the ideal solvent), $v \approx 0.5$. At the *theta* temperature, for a particular solvent, the polymer coil persists (above *theta* the coil undergoes expansion as it interacts with the solvent, while below *theta* the coil tends to collapse as the polymer segments attract one another, often leading to phase separation). The value of v increases as the solvent becomes better suited for dissolving the polymer and the deviation from the ideal state becomes greater.

Key Facts
● Polymer properties vary with molar mass and polydispersity $\overline{M}_w/\overline{M}_n$.

It is important to appreciate that the molar mass averages may give very different results for a polymer sample. For example, consider a hypothetical polymer sample consisting of chains of four discrete molar masses, 10 000, 20 000, 50 000, 100 000 g mol^{-1} in the ratio 1:3:6:1.

Using the equations mentioned earlier (2.3) and (2.4).

$$\overline{M}_n = \frac{(1 \times 10^4) + (3 \times 2 \times 10^4) + (6 \times 5 \times 10^4) + (1 \times 10^5)}{1 + 3 + 6 + 1}$$

$$= 42\,727\,\text{g mol}^{-1}$$

$$\overline{M}_w = \frac{[1 \times (10^4)^2] + [3 \times (2 \times 10^4)^2] + [6 \times (5 \times 10^4)^2] + [1 \times (10^5)^2]}{(1 \times 10^4) + (3 \times 2 \times 10^4) + (6 \times 5 \times 10^4) + (1 \times 10^5)}$$

$$= 55\,957\,\text{g mol}^{-1}$$

The two results are quite different, so it is important to quote the method by which a molar average has been calculated (or the analytical method from which it has been obtained). Unless the polymer is monodisperse, $\overline{M}_w$ is always greater than $\overline{M}_n$. There are other molar mass averages, but these are generally not commonly encountered.

Molar Mass Distribution (Polydispersity Index)

Having obtained both $\overline{M}_n$ and $\overline{M}_w$ (some techniques such as gel permeation chromatography, also known as size exclusion chromatography, give these data as a matter of course) it is possible to determine the polydispersity index (or *heterogeneity index*) of the polymer by calculating the ratio $\overline{M}_w/\overline{M}_n$. This gives a rough measure of the breadth of the molar mass distribution and as already noted may be of interest when examining the properties (or the processability) of a polymer. For many polymerisations the polydispersity index is around 2.0. As $\overline{M}_w$ is always greater than $\overline{M}_n$ (unless the sample is monodisperse) (Figure 2.16), then the polydispersity index is greater than unity.

As we have already said, a polymer can be manufactured with a range of properties and this is achieved by varying the average molar mass distribution, the polydispersity index, the degree of crystallinity (degree of branching, density), and the tacticity. A simple example to illustrate the effect of tailored properties is plasticity. The presence of low RMM chains does tend to soften (plasticise) the bulk polymer, while high RMM chains tend to raise the melt viscosity of a polymer (as a result of the greater opportunity for the chains to undergo entanglement). As most synthetic

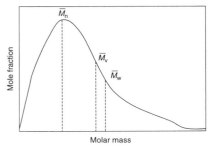

Figure 2.16 Typical molar mass distribution for a synthetic polymer sample showing $\overline{M}_n$, $\overline{M}_v$ and $\overline{M}_w$

polymers consist of a mixture of molecules of different lengths and composition, their properties depend on the molecular structure and composition of the mixture.

2.2.6 Immiscibility of Polymers

One beneficial aspect of the separation of waste streams is that, when mixed, many individual polymers are simply immiscible at a molecular level (due to differences in their molecular structures). This is explained in Textbox 2.3.

Mixtures that are known to be potentially immiscible (or only sparingly miscible) include PS/SAN, PS/ABS, PS/PVC, PS/PP, PS/LDPE, PS/HDPE, PET/PVC and PET/HDPE, etc.[1] Furthermore, when mixed, these polymers form heterogeneous or incompatible blends, unlike homogeneous blends (which are thermodynamically stable, do not undergo large-scale phase separation and exhibit a single glass transition temperature). However, in heterogeneous blends, many of the previously desirable mechanical properties of the separate homopolymers may be greatly diminished. Consequently, if separation of the waste stream is not undertaken, then the polymer waste mixture is only suitable for use in a relatively low cost secondary application. Lemmens[1] has discussed in detail the blending of polymers and the factors affecting (and facilitating) the miscibility of multi-component blends.

Key Facts
- Polymers are generally immiscible with each other and heterogeneous blends have inferior properties, so it is important to separate polymers in the waste stream.

Textbox 2.3 Why do plastics separate?

For a two-component polymer blend, phase separation into two phases (corresponding to the binodal compositions) is thermodynamically favoured, but an energy barrier must be overcome for this to occur.

Figure 2.17 shows schematically two types of phase boundaries commonly encountered in binary polymer blends and the third, more likely, behaviour. In the bottom example in (A), where the two-phase region is characterised by the upper critical solution temperature (UCST), the critical temperature (T_c) occurs near the maximum of the cloud-point curve. For most nonpolar polymer blends, ΔS_{mix} (the entropy of mixing) is normally positive, but influenced heavily by temperature and so the solubility (miscibility) of the polymers depends mainly on the magnitude of ΔH_{mix} (the enthalpy of mixing), which is normally endothermic (and hence positive). This means that, as temperature decreases, ΔG_{mix} (the Gibbs free energy of mixing) eventually becomes positive (the mixing process is not favoured) and phase separation of the polymer blend takes place.

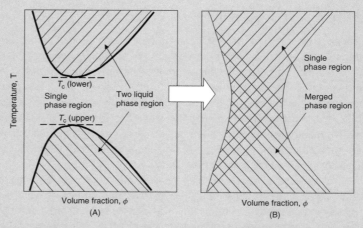

Figure 2.17 A schematic showing phase behaviour for binary polymer blends: (A) two-phase region showing upper critical solution temperature (UCST) and lower critical solution temperature (LCST); (B) commonly encountered phase separation behaviour in binary polymer blends

In contrast, in the top example in (A) the two-phase region characterised by the lower critical solution temperature (LCST) is somewhat more complex with the cloud-point curve inverted and the corresponding T_c located at the minimum of the miscibility curve. This means that in practice there is a decrease in miscibility as the temperature rises. This phenomenon is most commonly observed in polymer blends containing polar components and results from the formation of inter-chain attractions that enhance miscibility, for example (thermally labile) hydrogen bonds. A rise in the temperature reduces the numbers of bonds present and reduces the miscibility of the blend's components, eventually leading to phase separation. LCST is an entropically controlled phenomenon. The final diagram (B) shows the type of phase separation that occurs most frequently in binary polymer blends. While a deeper discussion of this topic is outside the scope of this book, the reader is directed to Chapter 8 of Cowie's excellent textbook (see the list for further reading).

Key Facts
● A compatibiliser ideally penetrates both phases of a blend and interacts with each of the components.

2.2.7 Improving the Compatibility of Polymers

Unless separation is achieved, either by manual selection or by one of the techniques described below, then in order to maintain the mechanical properties of a polymer blend in a second application it may become necessary to use a chemical compatibiliser. This is generally a relatively simple chemical (in structural terms) that has a tendency to exhibit phase separation and crucially is able to penetrate readily into the different phases of the blend and interact with each of the components (Figure 2.18). For example, block and graft copolymers have been found to improve the compatibility of mixtures of parent homopolymers, leading to a blend that exhibits a single T_g, despite being made up of more than one component. The copolymer can be thought of as acting like a 'macromolecular surfactant', promoting and stabilising the emulsion of the molten homopolymers. It is possible to tailor the structure of the copolymer to introduce either blocks of the main chain or grafts that can readily penetrate into the different phases. The molecular weight (chain length) may also be altered as optimal properties are obtained when the magnitudes of both homogeneous polymers

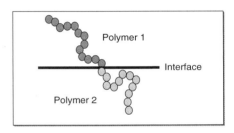

Figure 2.18 Idealised compatibilising block copolymer (shown distributed between the two polymer components of a blend)

and the compatibilisers are approximately equal. However, it is important that the compatibiliser is not too miscible with any of the homogenous polymers since this leads to a reduction in the strength of the bonding point at the boundary surface.

Often only small interactions may be sufficient to cause surface activity and improve the blend's mechanical properties. For example La Mantia[2] describes the use of a thermoplastic elastomer Kraton 1901X (a triblock copolymer of PS, partly hydrogenated polybutadiene/maleic anhydride) to enhance the compatibility of PET and PP (in an otherwise incompatible blend). For the unmodified PET/PP blend an elongation-at-break of 5 % was recorded and an impact strength of 27 J m^{-1}. The addition of 10 % Kraton 1901X improved these figures significantly (to 320 % and 123 J m^{-1} respectively). While the original PET/PP blend displayed dispersed PET particles within a continuous PP matrix, the addition of the compatibiliser led to a reduction in the dimensions of the PET particles and an improved adhesion between the components. A number of commercial compatibilisers are available and have been tailored specifically for use with recognised commercial polymer blends.

2.3 THE INFLUENCE OF THE ENVIRONMENT ON POLYMERS

In this section we look at the effect of the environment on polymers, in order to understand how polymer materials degrade and how that affects their useful lifetime and the implications for the subsequent waste management of polymers. In an ideal world, we would want our structural or insulating polymers to be able to maintain their pristine colour, strength and shape under all possible environmental conditions of heat, light, moisture, and exposure to aggressive gases; until that is, we want to dispose of them. Then we want to be able to separate, recycle or degrade them down to their elemental constituents. In the first instance, we want to inhibit degradation; in the second, we want to accelerate it. A polymer in use interacts with the environment and degrades slowly whatever we do to it and all polymers are susceptible to thermal, hydrolytic, oxidative or photolytic degradation to some degree or other, depending on their composition and the environment in which they are used. Unfortunately, most synthetic polymers are immune to natural biodegradation processes which makes their disposal much more difficult.

You only need to pick up the plastic water can that has been left out in the garden for a couple of years and spill its contents over your shoes as the handle breaks to see that the main effect of ageing is loss of strength. A cursory glance at the offending article will also reveal the secondary effect of ageing, *i.e.* discoloration. Both arise from chemical disintegration of the polymer structure, *i.e.* depolymerisation and destruction of hydrogen bonds or chemical cross-links between polymer chains.

Perhaps the most costly example of polymer degradation in recent times was the loss of a space shuttle, with its crew, which blew up as a result of hardening of elastomeric seals and consequent loss of fuel. Some recent plane crashes have been blamed on degradation of (polymeric) electrical insulation and consequent shorts in critical electrical circuits.

2.3.1 Mechanisms of Degradation/Depolymerisation

Prolonged exposure to light and heat (hot water) destroys cross-linking and causes yellowing or loss of glaze and eventually embrittlement. Polyethylene and more recently polypropylene have been extensively used as mouldable, pliable containers. However, both are susceptible to light, which induces depolymerisation and hardens and embrittles the material. Although modern production methods include the addition of stabilisers to decrease the rate of degradation, polyethylene medical containers, for instance, which have to be regularly sterilised, can still deteriorate rapidly. Another example is PVC, which in its natural form is highly susceptible to the detrimental effects of heat, light and moisture. It is also unprocessable, because it decomposes at 180 °C, just 20 °C above its softening temperature. Modern methods of production involve the use of copolymers, stabilisers and lubricant additions and have considerably extended the use of rigid PVC. We will look in more detail at the mechanisms of degradation and methods of stabilising these polymers later.

Polymer degradation has considerable influence on polymer recycling because it may affect a polymer's processability. As discussed later in the book, some polymers can be recycled mechanically several times without significant degradation of their properties; however, as the number of reprocessing cycles increases, the properties start to deteriorate progressively. Some polymers may also be suitable for recovery of monomers by chemical recycling. However, owing to the loss of properties by degradation, for many plastic materials the only feasible option at the end of their useful life is either incineration energy recovery or disposal in landfill.

2.3.2 Factors Affecting the Chemistry of Degradation

The chemistry of degradation is highly dependent on the chemical nature and type of the polymer and, as we will show in Chapter 3, there is a wide variety of polymers in use today. The susceptibility of a polymer to degradation depends on both its chemical and physical characteristics and those of its surroundings.

The following polymer properties are susceptible to degradation:

- Crystallinity: in general the amorphous regions of a polymer are more rapidly degraded than the crystalline regions, because there is a lower level of cross-linking between chains and because the structure is more open, allowing for easier air or aggressive gas access.

Key Facts
● All polymers degrade in the environment as a result of bond scission under the influence of heat, light, moisture or oxygen.
● Chemical and bulk properties of a polymer influence the rate at which it degrades.

- Glass transition temperature: above the transition temperature the polymer molecules are more mobile and more susceptible to certain types of degradation.
- Functional groups: certain types of side chain groups are susceptible to chemically degrading reactions, *e.g.* dehydration of –OH groups. Chromophoric side chains such as >C=O are sensitive to UV light. Tertiary hydrogen atoms are susceptible to attack in processes where free radicals are generated.

Degradation of these properties is accelerated by various environmental agents including:

- Heat: all chemical compounds decompose eventually when heated and polymers are no exception. Figure 2.19 shows the relative thermal stability of some common polymers as a function of the onset temperature for thermal degradation.

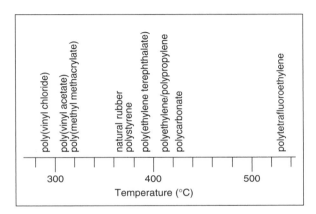

Figure 2.19 Relative thermal stability of some common polymers

Key Facts
- Gas access, backbone structure and the presence of impurities influence the rate of degradation.

- Light: the energy of the UV component of natural light is sufficiently high to break chemical bonds and polymers with chromophoric side chains are particularly sensitive.
- Moisture and oxygen are the main causes of chemical degradation by the oxidation/hydrolysis of side chain groups or direct attack on intermonomer linkages in the polymer chain. Other aggressive gases such as ozone and sulphur dioxide also have an important effect.
- Bio-active organisms: bio-degradation is a particularly important area of research at present (see Chapter 8). Most natural polymers are susceptible to fungal attack; most synthetics are not.

2.3.3 Mechanisms of Degradation

Three fundamental properties of polymers determine their stability:

- Access: molecules are held in a rigid structure in close proximity to each other and reaction depends on freedom of access of attacking species or on suitable side chains being in close enough proximity to each other to interact.
- The backbone structure: reaction can propagate along the backbone causing an 'unzipping' of the molecular structure.
- Impurities: whether deliberately or adventitiously added, impurities have a profound effect on the stability by accelerating or inhibiting the degradation reactions.

Three basic degradation mechanisms can be identified:

(i) scission of intermonomer linkages in the backbone;
(ii) scission of side chain linkages in the backbone;
(iii) ionically catalysed attack on side chains.

(i) Scission of intermonomer linkages in the backbone: Figure 2.20 shows an example of the thermal degradation of polystyrene. Products of degradation may be monomer units, or short chain oligomers of the monomer or products of degradation of the monomer.

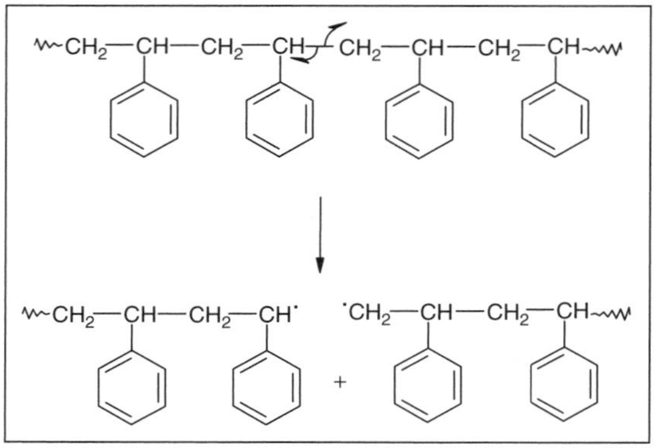

Figure 2.20 Scission of the polystyrene backbone during thermal degradation

Depolymerisation frequently occurs with the formation of radicals and is therefore initiated by radical formers and inhibited by radical scavengers. The general form of the mechanism of thermal degradation is often represented by the following equations:

$$
\begin{aligned}
\text{Random initiation} && M_n &\longrightarrow M_j^* + M_{n-j}^* \\
\text{Terminal initiation} && M_n &\longrightarrow M_{n-1}^* + M^* \\
\text{Depropagation} && M_i &\longrightarrow M_{i-1}^* + M \\
&& M_i &\longrightarrow M_{i-z}^* + M_z \\
\text{Intermolecular transfer} && M_i^* + M_n &\longrightarrow M_i + M_n^* \\
\text{Scission} && M_n^* &\longrightarrow M_j + M_{n-j}^* \\
\text{Termination} && M_i^* + M_j^* &\longrightarrow M_i + M_j \cdot M_{i+j}
\end{aligned}
$$

Where n, m, i, j are the number of monomers in a chain and z the number of monomers in a termination unit.

However, not all reactions occur by a radical mechanism. For instance, Figure 2.21 shows the mechanism of thermal degradation of poly(ethylene terephthalate), which involves the β-hydrogen and scission of the alkyl–oxygen bond *via* electron transfer around a six-membered ring.

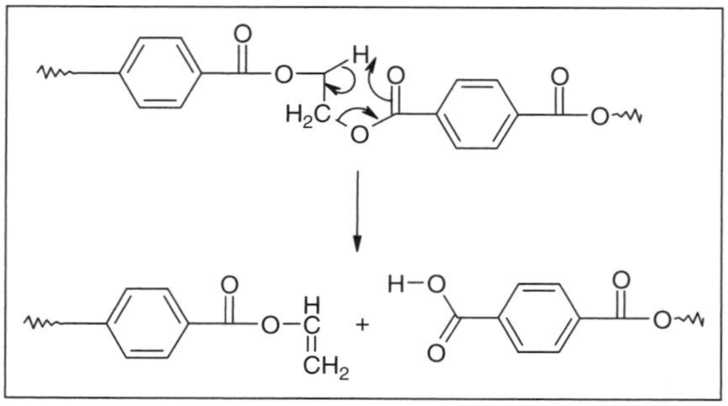

Figure 2.21 Mechanism of the thermal degradation of poly(ethylene terephthalate)

Polyolefins are particularly susceptible to attack by oxygen and undergo an auto-oxidative degradation, which can be represented as follows:

$$
\begin{aligned}
\text{Initiation} && RH + X^* &\longrightarrow R^* + XH \\
\text{Chain reaction} && R^* + O_2 &\longrightarrow ROO^* \\
&& ROO^* + RH &\longrightarrow ROOH + R^* \\
&& 2R^* &\longrightarrow R - R \\
\text{Termination} && ROO^* + R &\longrightarrow ROOR \\
&& ROO^* &\longrightarrow \text{Disproportionation products}
\end{aligned}
$$

In this case, an asterisk is used to indicate a free radical and all the reactions involve electron transfer processes, initiated by an adventitious free radical X*. The key processes are then the formation and reaction of peroxide radicals by reaction with oxygen. We will discuss these processes a little later and the strategies for interfering with them to extend polymer life.

(ii) Scission of side chain linkages to the backbone: Figure 2.22 shows the example of the photolytic degradation of poly(methyl acrylate) at ambient temperatures. In higher methacrylates, such as poly(tert-butyl methacrylate), cyclic degradation of the side group can occur by a mechanism similar to that of poly(ethylene terephthalate), as shown in Figure 2.23. Products of degradation will be related to the composition of the side group, although, in some cases, further reactions of the side chain can occur, such as the cyclisation of polyacrylonitrile shown in Figure 2.24.

CHAPTER 2

Key Facts
● Degradation occurs by free radical/electron transfer processes, and scission of side-chain linkages or the backbone by ionically catalysed attack.

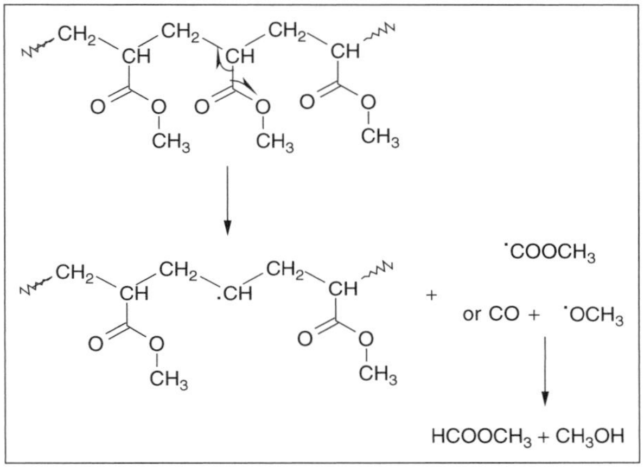

Figure 2.22 Photolytic degradation of poly(methyl acrylate)

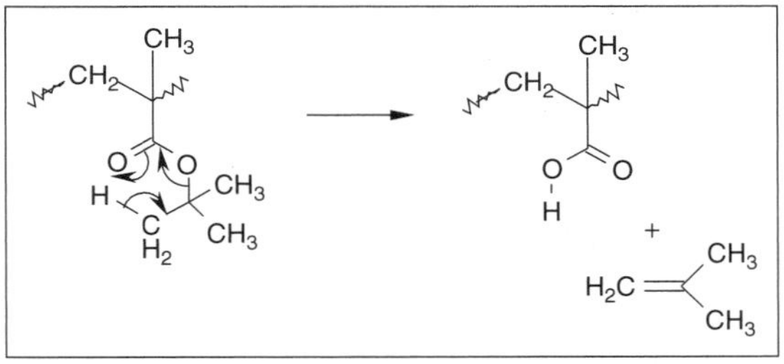

Figure 2.23 Cyclic degradation of side groups in poly(tert-butyl methacrylate)

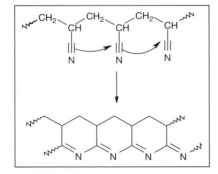

Figure 2.24 Cyclisation of side groups in polyacrylonitrile

35

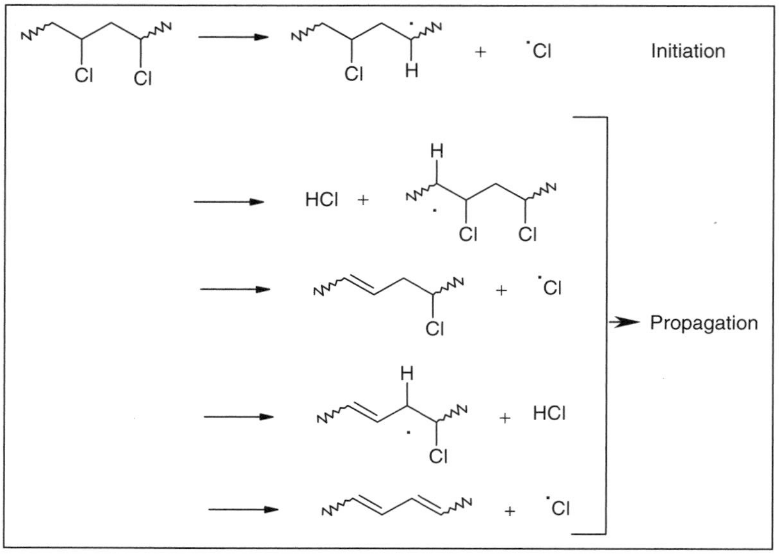

Figure 2.25 Possible mechanisms for the thermal degradation of PVC

A commercially important polymer, which degrades by loss of the side chain element is PVC. The overall degradation of PVC can be represented as:

$$\ldots CH_2-CHCl-CH_2-CHCl \longrightarrow \ldots CH=CH-CH=CH \ldots + HCl$$

but the precise mechanism of decomposition is not yet fully elucidated. Figure 2.25 shows possible routes for initiation and propagation.

The formation of double bonds leads to discoloration and reduction in its mechanical properties, and the low overall thermal stability of PVC is thought to be caused by the whole range of irregular structures in the polymer, *e.g.* initiator residues and internal unsaturation (C=C bonds). The temperature at which (gaseous) HCl emission first becomes detectable depends on the degradation conditions, but unstabilised PVC typically undergoes some HCl (<5% by weight) evolution at around 100 °C, followed by rapid thermal degradation at around 230–250 °C[3] and rapid evolution of HCl. Apparent activation energies for the thermal degradation of PVC range between 132.5 kJ mol^{-1} at 10% decomposition and 157.3 kJ mol^{-1} at 90% decomposition[4]. Traditionally, degradation has been inhibited by the addition of radical scavengers (see discussions of stabilisation and additives later in the chapter), but the HCl formed also autocatalyses the reaction. New degradation inhibitors are designed to scavenge HCl either as it is liberated (a primary stabiliser) or acting to prevent oxidative degradation by hindering the attack of oxygen on the chain (a secondary stabiliser). Substituted carbazoles are effective secondary stabilisers, significantly reducing the emission of HCl into the environment[4].

(iii) Ionically catalysed attack on side chains: Figure 2.26 shows the example of a proton-catalysed cyclisation of rubber, which causes cross-linking, hardening and loss of ductility.

2.3.4 Kinetics of Degradation/Depolymerisation

Degradation of Natural and Synthetic Straight Chain Polymers

Understanding the kinetics of degradation is important if we are to predict the useful life of a polymer. Kinetic equations for degradation were first derived in the mid-1930s by Kuhn[5], Ekenstam[6] and co-workers for the case of a linear polymer undergoing random degradation. They made the assumption that the rate of degradation at time *t* is proportional to the

Key Facts
● PVC discolours due to formation of unsaturated >C=C< bonds and decomposes above 230 °C with evolution of (gaseous) HCl.

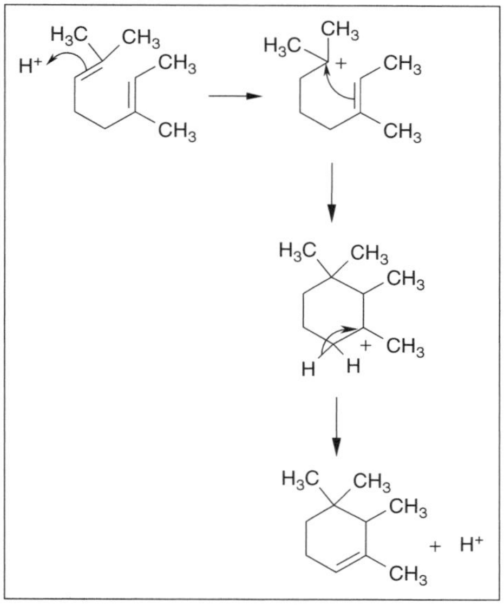

Figure 2.26 Ionically catalysed cyclisation of rubber

total number of unbroken inter-monomer bonds remaining at time t. This is equivalent to a first-order reaction mechanism in simple chemical terms and the derivation is given in most polymer text books. The resulting equation is quoted in terms of polymer DP as shown in Equation (2.7).

$$\log\left(1 - \frac{1}{DP_t}\right) - \log\left(1 - \frac{1}{DP_0}\right) = -kt \tag{2.7}$$

where DP_t and DP_0 are the DP at times t and 0 respectively. If DP_t and DP_0 are large then, mathematically, this simplifies to the zero order equation:

$$\frac{1}{DP_t} - \frac{1}{DP_0} = kt \tag{2.8}$$

This approach is strictly applicable only in the following circumstances:

- the polymer chain is linear and of high molecular weight;
- the polymer is mono-disperse and the products of scission are themselves long chain molecules;
- there is a low degree of chain end-chopping;
- there is no loss of monomer units during scission.

The Ekenstam equation (Equation 2.8) was derived from the assumption that all bonds in the polymer are the same and that degradation proceeds at random throughout the material. Conformance to the equation is therefore generally assumed to demonstrate a random chain scission mechanism and many linear polymers obey these simple kinetics over a wide range of conditions.

The 'order of reaction' of Equations (2.7) and (2.8) is often confused because the original assumptions for the derivation of Equation (2.7) are first-order. However, the approximation to Equation (2.8) makes the implied assumption that the number of bonds has not changed significantly and can therefore be assumed to be constant, which makes the reaction effectively zero order. In practice, there is no discernible difference between the two approaches at measurable DP values, as can be seen in Figure 2.27 for the degradation of cellulose the main constituent of paper and most nonplastic packaging. However, careful examination has revealed that the relationship is only truly linear in the early stages

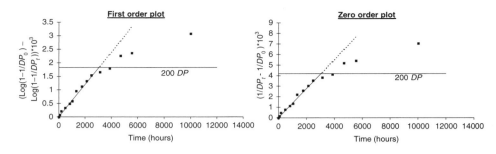

Figure 2.27 Deviation from the Ekenstam equation for cellulose ageing in air at $120\,°C$

Key Facts

● Many polymers
degrade randomly
with kinetics,
which, to a first
approximation, obey
a zero order model.
● Polymers are not
homogenous
materials so, in
reality, scission is
not truly random
and deviations from
simple kinetic
models occur.

of ageing. Deviations are observed at low *DP* values, even when plotted using the full first-order equation, as shown in Figure 2.27.

The reasons for this probably lie in the inhomogeneous nature of the material. Cellulose is a semi-crystalline material with areas of high crystallinity and others that are totally amorphous. It has a wide molecular weight range with individual molecules meandering between crystalline and amorphous regions. It is therefore highly unlikely that all bonds have the same probability of scission. If we assume that k in Equation (2.7) decreases as ageing proceeds by a typical first-order type process as the more reactive bonds are destroyed, such that:

$$k_1 = k_{1_0} e^{-k_2 t} \tag{2.9}$$

where k_0 and k_2 are constants, we can derive a modified version of the Ekenstam equation (Equation 2.10), which fits the data better[7], as shown in Figure 2.28.

$$\frac{1}{DP_t} - \frac{1}{DP_0} = \frac{k_{1_0}}{k_2}(1 - e^{-k_2 t}) \tag{2.10}$$

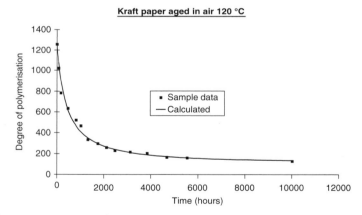

Figure 2.28 The same cellulose data (as in Figure 2.27) fitted assuming that the rate constant decreases with time

Degradation of Thermoplastics, Cross-Linked, Blended and Filled Polymers

The degradation kinetics of most commercially important polymers is extremely complex and outside the scope of this book. Suffice it to say that filled and blended polymers, and to some extent cross-linked polymers, are heterogeneous materials with one phase dispersed within another. The presence of cross-linkages stabilises polymers towards degradation and degradation rates need to be measured separately for each polymer and can vary with a single polymer type, depending on the degree of cross-linking. In blended polymers,

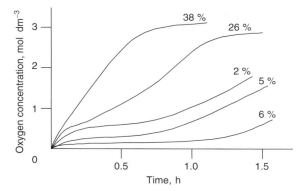

Figure 2.29 Oxygen uptake by carbon-filled polyethylene at 2 %, 5 %, 6 %, 29 % and 38 % by weight

the degradation of each phase might be expected to proceed independently, whereas, in the case of filled polymers, account must be taken of interactions with the filler at the polymer–solid surface interface. The introduction of fillers may inhibit or accelerate degradation. An example is the inhibition of oxygen uptake by carbon in polyethylene as shown in Figure 2.29, but note also that too much carbon increases the rate of carbon uptake and defeats the object of the filler.

2.3.5 Consequences of Degradation

The Kuhn and Ekenstam analysis of the degradation of linear polymers by random chain scission shows that the molecular weight decreases as degradation proceeds, as the polymer chains are broken down to smaller units by scission of bonds in the backbone (depolymerisation). Under more vigorous conditions of degradation, vaporisation may also occur. The main effect of depolymerisation is loss of mechanical strength resulting from:

- Loss of cross-linking, which may arise as a result of scission of direct inter-monomer bonds or loss of hydrogen bonding.
- Reduced chain length reducing the number of chemical and physical (*e.g.* entanglements) chain interactions.
- Loss of tensile strength with decreasing *DP*, although some polymers, such as starch-filled polyethylene, undergo a small increase in tensile strength in the very early stages of degradation, due to increased cross-linking as cleaved molecules relax into more stable configurations.

Other effects of depolymerisation are yellowing and loss of optical properties generally and hardening, which results from increased cross-linking as broken bonds re-form between adjacent molecules.

2.3.6 Stabilisation/Destabilisation by Additives

Most modern polymers contain stabilising (or occasionally destabilising) additives. The former are used to reduce the environmental effects on polymer properties and increase their lifetime, while the latter can help in degradation of waste polymers into less persistent constituents thus reducing the effect polymers have on the environment.

Additives can be categorised as:

- processing aids (to improve processibility);
- antioxidants (or stabilisers);
- mineral fillers (to bulk out the polymer);
- impact modifiers (*e.g.* glass fibres to increase strength);
- compatibilisers (to improve mixing of two or more polymers).

Key Facts
- Rates of degradation are strongly influenced by cross-linking, blending and presence of fillers.
- Chain scission, due to degradation (depolymerisation) leads to loss of mechanical strength and yellowing.

Key Facts ⚹━━━◁
● 'Additive'
stabilisers, in the
form of a dispersed
phase, tend to leach
out during
recycling, causing
loss of stability.

Two general types of additive are used in polymers, namely the 'additive' type and the 'reactive' type. Additive type stabilisers are generally physically incorporated polymers as a dispersed phase, because it is simple and the most economical way of introducing the material. However, this leads to a variety of problems, such as poor compatibility and leaching (into the environment) particularly during recycling, which leads to loss of mechanical properties.

The application of reactive additives is exemplified by new approaches to flame retardancy. It involves either the design of new polymers, e.g. intrinsically flame retarding polymers, or the modification of existing polymers through copolymerisation with an inhibiting unit, either in the chain or as a pendant group. Currently, new polymer designs lack sufficient versatility in manufacturing and processing or are uneconomical, which leaves the modification approach as the most favoured method of protection. For instance, by incorporating P-, Si-, B- or N-bearing units directly (via covalent bonds) into the polymer backbone, it is possible to impart permanent flame retardancy, without reducing the original physical and mechanical properties of the polymer[8].

Three commercially important cases will be considered: the stabilisation of polyolefins; the use of reactive ligands to destabilise polymers for biodegradability and the addition of flame retardants to polymers.

Stabilisation of Polyolefins

The auto-oxidation mechanism of degradation of polyolefins outlined earlier in the chapter can be extended to include the effects of UV light. The resulting mechanism is redrawn schematically as two interlocking cycles in Figure 2.30. Cycle A is the auto-oxidative reaction involving alkyl and alkyl peroxide radicals, which is fed by oxygen and terminated by acid formation. The photolytic component in cycle B arises when UV light decomposes the acid products of cycle A into hydroperoxides.

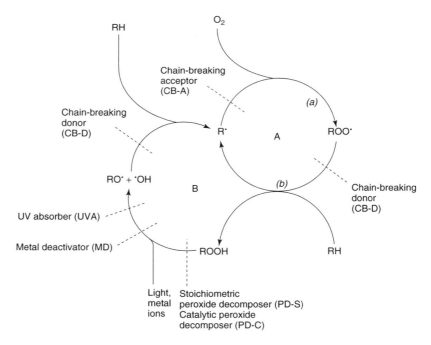

Figure 2.30 Auto-oxidative and photolytic degradation cycles of polyolefins

They decompose to alkyl oxide and hydroxide radicals and feed the decomposition cycle B. The principal points where the chain reactions can be interrupted, either by radical donors or acceptors or by UV absorbers, are indicated in the figure. Electron acceptors can oxidise the alkyl radical to a carbonium ion and subsequently to an inert olefin by loss of a proton. Alternatively, the alkyl peroxide can be reduced to a hydroperoxide ion and subsequently to an acid, as shown in Figure 2.31. The oxidation of alkyl radicals to an olefin happens in direct competition with the reaction with oxygen in cycle A and is only successful when the oxygen concentration at the oxidation site is low.

Key Facts
● Polyolefins degrade *via* alkyl oxide or hydroxide radicals and are stabilised by the addition of radical donors or acceptors or by UV absorbers, such as hindered amine light absorbers HALs.

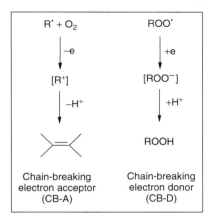

Figure 2.31 Alkyl radical oxidation and hydroperoxide ion reduction by polyolefin anti-oxidants

The degradation of polyolefins (and other polymers) can be retarded by addition of different stabilisers. Most stabilisers for polymers are mixtures of a number of anti-oxidants, which operate by similar, complementary mechanisms and frequently complement and enhance each other's effectiveness (often described as 'synergistic enhancement'). For instance, an anti-oxidant operating in the B cycle on the hydroperoxides reduces the number of alkyl radicals in the A cycle and thereby enhances the effects of a second anti-oxidant acting in the A cycle and *vice versa*.

Typical electron-donating anti-oxidants are sterically hindered phenols, with electron-releasing side groups to stabilise the transition state and increase activity. Two examples are shown in Figures 2.32 and 2.33.

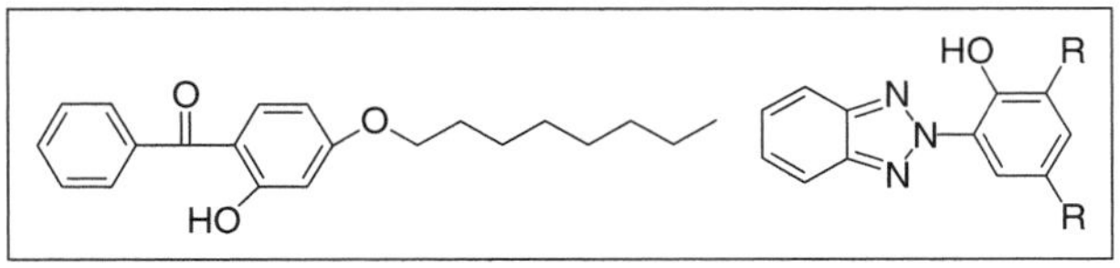

Figure 2.32 Sterically hindered phenolic inhibitors for light stabilisation

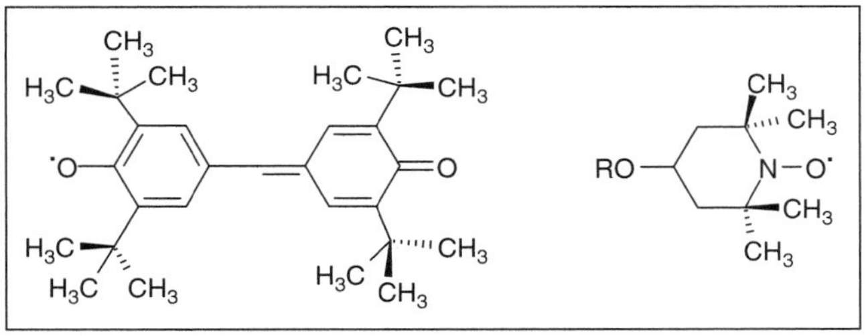

Figure 2.33 Galvinoxyl and nitroxyl radical

Typically oxidising, anti-oxidants are 'stable' radicals, such as the following galvinoxyl and nitroxyl radicals, that stabilise the charge by distributing it over conjugated or cyclic ring systems.

41

Geuskens *et al.*[9] have show in accelerated laboratory tests that hindered amines can extend the useful life of hydrocarbon polymers by:

(a) increasing the induction time prior to the onset of degradation by a factor of 15 at a concentration of 0.5 % and 25 at 1 %;
(b) decreasing the rate of degradation once it does start to half (at 0.5 %) and to a less than a quarter (at 1 %) of the uninhibited rate.

Flame Retardants and other additives in Polymers

Key Facts 🔑
● Fillers reduce flammability by diluting the polymer or releasing non-flammable gases.
● Phosphorus promotes char formation and intumescent materials swell, reducing gas excess to the flame.
● Many flame retardants produce toxic or carcinogenic products during combustion.

Fire hazards associated with the use of these polymeric materials, which can cause the loss of life and property, are of increasing concern among government regulatory bodies, consumers and manufacturers. The use of flame retardants to reduce combustibility of the polymers, and smoke or toxic fume production, therefore becomes a pivotal part of the development and application of new materials. The construction, electrical/ electronic components and transportation industries are major markets where flame retardants are required.

All flame retardants act either in the vapour phase or the condensed phase through a chemical and/or physical mechanism to interfere with the combustion process during heating, pyrolysis, ignition or flame spread. For example, the incorporation of fillers mainly acts to dilute the polymer and reduce the concentration of decomposition gases. Hydrated fillers also release nonflammable gases or decompose endothermically to cool the pyrolysis zone at the combustion surface. Halogens, phosphorus and antimony act in the vapour phase by a radical mechanism to interrupt the exothermic processes and suppress combustion. Phosphorus can also act in the condensed phase promoting char formation on the surface, which acts as a barrier to inhibit gaseous products from diffusing to the flame and to shield the polymer surface from heat and air. Intumescent materials swell when exposed to fire or heat to form a porous foamed mass, usually carbonaceous, which in turn acts as a barrier to heat, air and pyrolysis products.

However, the benefits of flame retardants are muted by their potential health and environmental impacts. Halogen-based flame retardants for instance can produce toxic substances during combustion, *e.g.* toxic hydrogen halide. Furthermore, the EC proposes to restrict the use of brominated diphenyl oxide flame retardants, because highly toxic and potentially carcinogenic brominated furans and dioxins may form during combustion. Dioxins, together with some other additives, such as phthalates used in PVC toys, are also classified as endocrine disruptors, a diverse group of synthetic and naturally occurring chemicals that have the potential to affect adversely the health and reproductive fitness of humans and wildlife[10]. Many other chemicals, which either exhibit oestrogenic effects or interact with some part of the endocrine system, include organochlorine pesticides, polychlorinated biphenyls (PCBs), phenolic compounds, phthalate esters, dioxins and furans, alkyl ethoxylates and heavy metals. Unfortunately, these molecules do not degrade easily in the environment and may enter the air and/or aquatic environment and also the food chain. Some of the materials (*e.g.* phthalate esters) are used as processing aids in polymer production and their fate during recycling may be worthy of further investigation.

2.4 SUMMARY AND LEARNING OUTCOMES

Polymers form a unique family of materials with characteristics that make them of great technological importance in the 21st century. The field is also still growing rapidly as the skills of the synthetic polymer chemist are employed to yield monomers and polymers with physical and mechanical properties that are increasingly superior to those of traditional structural materials (*e.g.* wood, glass and metal). Armed with a knowledge of the subtle effects of molecular structure and process conditions on the resulting polymer and its properties, it is possible to tailor the product for specific applications. This may be in response to the particular requirements of the application in demanding very specific mechanical or optical properties, or to design polymers to be more 'user friendly' at the end of their working lifetime. We will examine selected important commercial polymers in the next chapter.

The mechanisms and kinetics of polymer degradation are as complex as the polymers themselves. It is not, therefore, surprising that they cannot be described simply, nor that they are not yet fully understood. There are, however, common elements in the patterns of degradation, which can be used to simplify our picture of the processes involved. The involvement of radicals in the depolymerisation reaction, for instance, allows us to design additives to inhibit degradation. The relatively simple kinetic description of random chain scission reactions of the macromolecules allow us to model rates of degradation under accelerated conditions and to extrapolate them to real conditions to predict durability of polymer materials.

By reading the contents of this chapter and attempting the revision exercises, you should be able to discuss:

● The basic principles of polymer chemistry, including nomenclature and basic definitions used in polymer science.
● The methods employed to produce polymers *via* step growth and chain growth routes and appropriate examples of commodity polymers to illustrate these processes.
● The differences between thermoplastic and thermosetting polymers, including polymer morphology and stereochemistry. The glass transition is of particular importance for a variety of reasons and you should be able to discuss the molecular features that influence its magnitude.
● The way in which the polymer's average molar masses can be calculated, their relation to the polydispersity index and the implications for polymer processing and/or recycling.
● The external conditions and internal structures that influence degradation of polymers.
● The different chemical mechanisms by with polymers degrade and the chemical and mechanical consequences of degradation.
● The kinetics of degradation of straight chain polymers, which can be used to predict the durability of a material under given environment conditions.
● The general techniques that can be employed to protect polymers from degradation and the structures and reaction mechanisms of common inhibitors.

Notation

DP	Average degree of polymerisation
DP_0	Initial average degree of polymerisation (time $=$ zero)
DP_t	Average degree of polymerisation after time t
K_v	Mark–Houwink constant
M_i	RMM of individual polymer chain
$\overline{M}_n$	Number average molar mass
m_0	Monomer molecular weight
$\overline{M}_v$	Viscosity average molar mass
$\overline{M}_w$	Weight average molar mass
N_i	Number of polymer chains (molecules) with an RMM of M_i
T_c	Critical temperature
T_g	Glass transition temperature
T_{gc}	Glass transition temperature of a copolymer
T_m	Melting temperature
V	Mark–Houwink constant
ΔG_{mix}	Gibbs' free energy of mixing
ΔH_{mix}	Enthalpy of mixing
ΔS_{mix}	Entropy of mixing
$[\eta]$	Limiting viscosity number (intrinsic viscosity)

2.5 REFERENCES AND FURTHER READING

1. Lemmens, J. (1995). *Compatibilizers for plastics*. In *Recycling and Recovery of Plastics*, eds. Brandrup, J., Bittner, M., Menges, G. and Michaeli, W., Hanser, Munich, pp. 315–326.

2. La Mantia, F.P. (1998). The role of additives in the recycling of polymers, *Macromol. Symp.*, **135**, 157–165.

3. Hamerton, I. Pielichowski, J. and Pielichowski, K. (1994). A study of the thermal degradation of poly(vinyl chloride) in the presence of carbazole and potassium carbazole using TGA/FTIR, *Polymer*, **35**, 336–338.

4. Pielichowski, K. Hamerton, I., Pielichowski, J. and Stanczyk, P. (1998). A study of the thermal properties of poly(vinyl chloride) with novel epoxypropanecarbazole-based dyes by TGA/FTIR, *Eur. Polym. J.*, **34**, 653–657.

5. Kuhn, W. (1930). On the kinetics of depolymerisation of high molecular chains, *Ber.*, **63**, 1503.

6. Ekenstam, A. (1936). The behaviour of cellulose in mineral acid solutions: Kinetic study of the decomposition of cellulose in acid solutions, *Ber.*, **69**, 553.

7. Emsley, A.M., Heywood, R.J., Eley, C. and Ali, M. (1997). On the kinetics of the degradation of cellulose, *Cellulose*, **4**, 1–5.

8. Lu, S.Y. and Hamerton, I. (2002). Recent developments in the chemistry of halogen-free flame retardant polymers, *Prog. Polym. Sci.*, **27**, 1661–1712.

9. Gueskens, G., Kanda, M.N. and Nedelkos, (1994). A comparative study of the efficiency of typical hindered amine stabilisers in different hydrocarbon polymers, *Int., J. Polym., Mater.*, **24**, 19–29.

10. http://www.bch.msu.edu/labs/tz/endocrine.htm, accessed 9 July 2001.

Further Reading

J.M.G. Cowie, Polymers: Chemistry & Physics of Modern Materials, 2nd ed., Blackie, Glasgow, 1991.

P. Munk, Introduction to Macromolecular Science, John Wiley & Sons, New York, 1989.

K.J. Saunders, Organic Polymer Chemistry, 2nd ed., Chapman & Hall, London, 1988.

M.P. Stevens, Polymer Chemistry. An Introduction, 3rd ed., Oxford University Press, Oxford, 1999.

D. Watson and P. Lorimer, Polymers, Oxford Chemistry Primers Series, No. 85, Oxford Sci. Publ., Oxford, 2000.

R.J. Young and P.A. Lovell, Introduction to Polymers, 2nd ed., Chapman and Hall, London, 1991.

Allen, N.S., Chirnis–Padron, A. and Henman, T.J. (1985). The photo-stabilisation of polypropylene – A review, *Polym. Degrad. Stab.*, **13**, 31–76.

Grassie, N. 1956. *Chemistry of High Polymer Degradation Processes*, Butterworths Scientific, London.

http://www.chemheritage.org/Polymers+People/PREFACE.html (for information about both history and historical figures in polymer science).

http://mat.ethz.ch/d-werk/suter/bt_frames.html (for a basic glossary of IUPAC terms in polymer science).

http://chemistry.miningco.com/science/chemistry/cs/polymerchemistry/index.htm.

2.6 REVISION EXERCISES

1. For each of the following list of key terms used in this chapter write a concise definition in a polymeric context and illustrate it with an example:

   (a) amorphous   (c) copolymerisation
   (b) chain polymerisation   (d) constitutional repeat unit

C H A P T E R   2

(e) cross-linking
(f) degree of polymerisation
(g) elastomers
(h) free volume
(i) glass transition temperature
(j) homochain polymers
(k) intrinsic viscosity
(l) number average molar mass
(m) oligomer
(n) plasticiser
(o) polydispersity index
(p) polyalkenes/polyolefins
(q) recyclate
(r) step growth polymerisation
(s) stereochemistry
(t) stereospecific catalysts
(u) tacticity
(v) thermoplastic
(w) thermoset
(x) *theta* solvent
(y) viscosity average molar mass
(z) weight average molar mass

2. It is often quite difficult to produce a very high molecular weight polymer using step growth polymerisation methods. Considering the differences between step growth and chain growth methods, can you give reasons for this observation?

3. A hypothetical polymer sample contains five discrete molar masses, 5 000, 15 000, 30 000, 45 000, 90 000 g mol^{-1} in the ratio 1:2:4:6:1. Calculate the number average molar mass, the weight average molar mass and hence the polydispersity index of the polymer.

4. Plot the values of melting point (T_m) against degree of polymerisation (m) for the following series of polymers.

Constitutional repeat unit	T_m (K)				
	$m = 2$	$m = 3$	$m = 4$	$m = 5$	$m = 6$
$-[-(CH_2)_m-]-$	400				
$-[-(CH_2)_mCO \cdot O-]-$	395	335	329	335	325
$-[-(CH_2)_m- CO \cdot NH-]-$	598	538	532	496	506
$-[-(CH_2)_mCH_2- SO_2-]-$	573	544	516	493	–

Comment on the structural features that lead to the variations in T_m with reference to polyethylene.

5. Assuming that the following substituents are all of a similar size (*i.e.* they all occupy a similar volume), how do you think that T_g varies for the following series of substituted vinyl polymers?

$$+CH_2-\underset{\underset{R}{|}}{CH}\underset{n}{+}$$

R = OH or CH$_3$ or Cl

6. A synthetic elastomer (BUNA-S or SBR rubber), which exhibits both good mechanical properties and flexibility, can be formed by copolymerising polystyrene and poly(buta-1,3-diene). The glass transition temperature of a polystyrene homopolymer is recorded as 373 K, while that of a poly(buta-1,3-diene) homopolymer is 203 K. What are the values of T_{gc} for a copolymer containing (a) 52 wt% styrene (48 wt% buta-1,3-diene) and (b) 48 wt% styrene (52 wt% buta-1,3-diene)? What can you say about the physical nature of the copolymers when in use at 288 K?

7. Discuss some of the general mechanisms of degradation of polymers using some of the more common polymers as examples. Explain why some polymers harden when exposed to heat, whereas others discolour and/or become brittle.

45

8. The diagram in Figure 2.30 is often used to describe the process of polymer degradation. Using the diagram or other aids,
   (a) describe in general terms the mechanism of radical-induced degradation of polymers, and
   (b) describe briefly the general mechanism of degradation observed in real polymers, with some specific examples.

9. (a) Explain the use of inhibitors to interfere with the mechanisms of radical-induced degradation (using the above diagram or other aids). Give a simple example.
   (b) Discuss other approaches to the inhibition of polymer degradation, with examples.

10. List the primary influences on polymers that lead to degradation and loss of properties. Describe the mechanisms of ageing related to each influence and discuss the effects on polymers in use.

11. Write down the first-order equation for the kinetics of ageing of polymers.
    Describe briefly how it is generated including a mathematical derivation. The equation is more generally used as a zero-order approximation. Give the form of the zero-order equation and discuss the chemical implications of using the equation in this form.

12. (a) Name the most important additives and say what they are used for.
    (b) Which ones have a potential to cause environmental problems? Explain the mechanisms by which that can happen and support your explanations with examples.

13. Immiscibility is not always a problem; discuss the conditions under which immiscible polymers blends can have useful properties.

feeding the waste streams: sources of polymers in the environment

Chapter 3 – Circe Invidiosa (JW Waterhouse, 1892). Having fallen in love with a beautiful nymph called Scylla, the merman Glaucus begged the witch Circe to concoct a love potion to make Scylla return his affections. However, in typically tangled mythological fashion, Circe was also in love with Glaucus and instead formulated a poison to despatch her rival. Pouring the potion into a fountain, Circe watched Scylla step into the water to swim only to begin a monstrous meta-morphosis. The comparison with feeding the waste streams is an apt one: consider the effect of waste polymers on the environment, including the effect of endocrine disruptors on aquatic species. . .

3.1 INTRODUCTION

Before we discuss or assess the feasibility of recycling, it is essential to identify the source (*i.e.* the route through which end-of-life material enters the waste stream) and, more importantly, the chemical nature of the plastics waste. The post-consumer, commingled plastics waste is quite unlike the original discrete polymeric products. It represents a poorly characterised mixture of a wide variety of polymers, varying not only in terms of their form (*i.e.* size and shape), but also their type, physical characteristics (*i.e.* age, solubility, processability or thermal properties) and source of origin (*i.e.* the original manufacturer and production process).

However, in addition to describing the nature and characteristics of polymers at the end of their useful life, it is also important to understand what happens at the beginning of their life cycle, so that the appropriate recycling options can be chosen for their subsequent life cycles. In the following sections we therefore first concentrate on the beginnings of the life cycle and examine different routes for polymer production. This is followed by an overview of polymer consumption in different sectors and the possibilities for their recovery from the waste streams. Given their commercial importance and production volumes, the emphasis is inevitably on thermoplastic materials; however, some of the other polymers will also be discussed.

3.2 POLYMER PRODUCTION

As we have already discussed in Chapters 1 and 2, the majority of commodity plastics are derived from gas or from crude oil which, after processing and refining, generates monomers that are then used in the subsequent manufacture of polymers. Various processing routes can be used to obtain these polymers. Some of these, used for the production of major thermoplastic materials, are shown in Figure 3.1 and discussed in more detail in the following sections. Note that the figure does not show by-products and ancillary operations of the production process, nor some of the other routes by which these polymers can be produced.

A summary of the major primary and secondary uses of the polymers that will be discussed in this chapter can be found in Table 3.1. For a more in-depth discussion of the many industrial polymers used in everyday applications, you may wish to refer to *Organic Polymer Chemistry* by Saunders, who presents an excellent discussion of this topic (see the list for further reading).

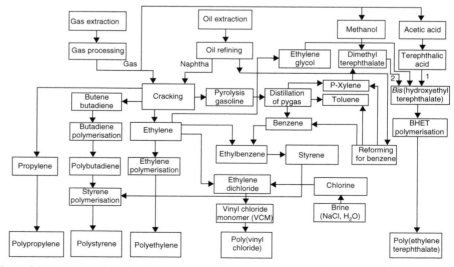

Figure 3.1 Conventional production routes for polymers from raw materials to commodity products (1, direct esterification; 2, ester interchange)

Table 3.1 Major commercial polymers and their primary and secondary uses.

	Thermoplastics	
	Primary use	Secondary use
PET or PETE Poly(ethylene terephthalate)	PET, a commonly recycled household plastic material, represents approximately 30 % of the plastic bottle market and is used to package a wide variety of food and beverage products such as soft drinks, juices, edible oils, liquor and butter. PET is valued for its clarity, toughness and ability to resist permeation by carbon dioxide.	Products made from recycled PET include carpets, insulating material in garments and sleeping bags (fibre fill), bottles, and containers, scouring pads, auto parts, tyres, paint brushes and geotextiles, such as landfill liners and industrial paints.
HDPE High-density polyethylene	HDPE is a characterised by its rigidity, low cost, ease of forming (e.g. injection moulding), and resistance to breakage. It is used to bottle milk, water, juices, bleach, detergents, motor oil, etc. For more than 20 years, HDPE has been the preferred material for drinking water and sewerage pipes. Now it is also used for gas pipes. The manufacture of HDPE packaging film is growing fastest and HDPE currently makes up around 21 % of the packaging market.	Recycled products made from HDPE include detergent and engine oil bottles, dustbins, recycling bins, soft drink bottle base cups, drainage pipes, animal pens, drums and pails, matting, milk bottle carriers, industrial pallets, plastic lumber, traffic barriers cones, flower pots, golf bag liners, kitchen drain boards and hair combs.
PVC Poly(vinyl chloride)/vinyl	Because of its blending capability, PVC (also commonly called vinyl) can be used to manufacture commercial products including heavy-walled pressure pipes, windows, doors, as well as crystal-clear food packaging, particularly bottles for cooking oils, water, household chemicals, food wraps, and health and beauty aids. Its properties include good clarity and chemical resistance.	Recycled PVC can be used to make drainage pipes, fencing, handrails, house siding, window frames, tiles, sewer pipes, traffic cones, garden and house drains.
LDPE Low-density polyethylene	LDPE is widely used in applications requiring clarity and processing ease. Its most common use is a film (73 % of consumption in Western Europe), for sacks, shrink wrap, stretch wrap, and refuse bags. LDPE is the predominant material for telecommunications cables and forms a water-vapour resistant seal in laminating of co-extruded multilayer films.	When recycled, LDPE can be used to make most of the products made from virgin LDPE (except for packaging).
PP Polypropylene	PP is resistant to chemicals, heat and fatigue. Consequently, it is widely used in many applications ranging from the manufacture of fibres and films to food packaging such as screw-on caps and lids, some yoghurt and margarine tubs, juice bottles and drinking straws. Polypropylene has long been used as the primary material for manufacture of automotive battery cases because it is lightweight, durable and cost-effective.	Products made from recycled PP, as well as potential markets for it, include auto parts, new automotive battery cases, bird feeders, furniture, buckets, water meter boxes, bag dispensers, golf equipment, carpets, refuse and recycling containers, grocery trolley handles and industrial fibres.

(continued overleaf)

49

Table 3.1 (*continued*)

Thermosets

	Primary use	Secondary use
PU Polyurethane	PU is a versatile material characterised by its toughness. It can be used for coatings, finishes, additives to improve resistance to chemicals and ozone, bumpers, gears, furnishing and car windscreen interlayers.	PU can be recycled into the same products as virgin PU (except for the applications which require high specification such as car windscreen interlayers) or different products (*e.g.* carpet underlay)
Epoxy resins (glycidyl ethers, see below for structure)	Epoxy resins of the glycidyl ether type are derived from epichlorohydrin and a polyphenol (usually bisphenol-A). They are widely used as structural plastics, surface coatings, adhesives, automotive components, electrical/electronic components, sports equipment and boats. They are characterised by low shrinkage on polymerisation, good adhesion, mechanical and electrical strength and chemical resistance.	There are few examples of recycling of epoxy resins, but those cited examine the use of nitric acid corrosion, energy recovery and particulation by regrinding.

Composites

	This comprises high performance polymers, frequently in the form of composites that are used in automotive and other durable goods applications. Composites are a combination of a polymeric matrix resin with fibre reinforcement and/or fillers. These polymers represent a significant recycling challenge because products made from the recycled mixed material tend to have poor physical properties; they are often brittle as most polymers are incompatible and do not chemically adhere to each other. Generally, the greater number of polymeric components in a blend, the poorer are its properties as recyclate.	Few products are currently being manufactured from recycled multi-material plastics, although limited ways do exist to improve the mechanical properties of products made from mixed plastics.

3.2.1 Polyethylene (PE)

Polyethylene is the simplest of the commercial polymers and forms the most widely commercialised group of the poly(1-alkene)s. Simplistically, it is formed by opening the double bonds of ethylene molecules and linking them together in linear or branched chains. Depending on how the polymerisation is carried out, the polymer chains may be highly linear or side-branched. The type of chains and the degree of branching determine the ultimate properties of the polymer, some of which are listed in Table 3.2 for the family of PE materials. The properties may be tailored by adjusting the polymerisation method or reaction conditions (see Figure 3.2) to favour differing degrees and forms of branching along the linear chain. The polymer chain length and degree of crystallinity (and hence the mechanical properties of the polymer) can be controlled by adding specific

Key Facts
● Polyethylene is the most widely commercialised poly(1-alkene).
● High density products have mainly linear chains whereas low density products have a high degree of branching.

Table 3.2 Characteristics of different forms of polyethylene.

Polyethylene	Density (g cm^{-3})	Number of branches (per 1000 carbon atoms)	Degree of crystallinity (%)	Additional comment
LDPE low density polyethylene	0.910–0.925	20–30 (methyl) 3–5 (n-butyl)	40–50	n-Butyl branches arise from 'backbiting'
VLDPE very low density polyethylene	0.890–0.915	Numerous	–	–
LLDPE linear low density polyethylene	0.910–0.925	–	–	Only contains short branches
HDPE high density polyethylene	0.942–0.965	<4 (Phillips) 5–7 (Ziegler)	60–80	No ethyl or butyl (Phillips) Ethyl branches (Ziegler)
MDPE medium density polyethylene	0.926–0.940	4–6	–	Formed from blending LDPE/HDPE or LLDPE copolymer

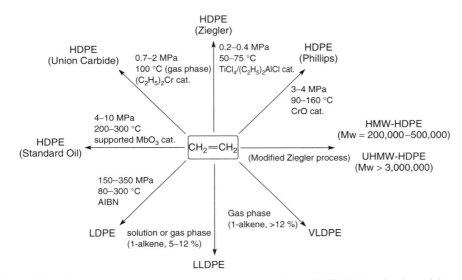

Figure 3.2 Different preparative routes to commercial polyethylene (HMW, high molecular weight; UHMW, ultra-high molecular weight)

amounts of hydrogen to the reactor. For example, linear chains, or chains bearing either few side chains or short side chains, pack more closely together and produce a high density, crystalline polymer with superior mechanical properties[1]. Heavily-branched chains, on the other hand, inhibit close packing and produce low density polymers with a high amorphous content. Other properties, such as impact strength (see Table 3.3), notched-impact strength and environmental stress cracking resistance all improve with increasing molecular weight, but this is accompanied by a reduction in the ease of processing.

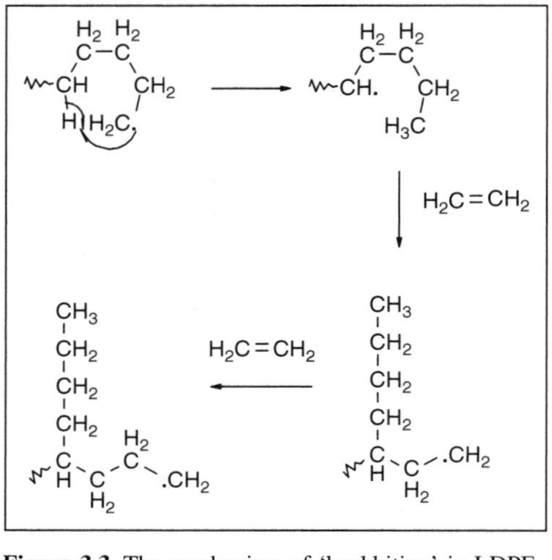

The following sections outline the production routes for different types of PE.

Low density polyethylene (LDPE), also known as 'linear' or 'rigid' PE, contains a high level of side branching with long side branches. It is prepared via free radical polymerisation of ethylene at relatively high temperatures and high pressures, depending on the choice of polymerisation initiator. Azodiisobutryonitrile (AIBN), benzoyl peroxide or oxygen are all used as initiators. The process is extremely exothermic, necessitating extreme caution with heat dissipation and control of the reactor atmosphere, which might form an explosive mixture of carbon, hydrogen and methane (see the aforementioned book by Saunders, *Organic Polymer Chemistry*, p. 49). As shown in Figure 3.3, the final polymer comprises both methyl groups and butyl residues arising from intramolecular transfer ('backbiting').

Figure 3.3 The mechanism of 'backbiting' in LDPE to produce butyl branches

Table 3.3 Some properties of selected commercial polymers.

Polymer	Additional comments	Density (g cm^{-3})	T_m (°C)[a]	Softening point (Vicat, °C)	Tensile strength (MPa)	Tensile strength (N m^{-2})	Elongation at break (%)	Hardness (Shore D)	Impact strength (Izod J m^{-1})
PE	LDPE	0.92	108	95	10	10 343	450	45	–
	LLDPE	0.94	123	–	10	10 343	700	55	–
	HDPE	0.96	133	120	28	27 580	500	65	–
PP	Homopolymer	0.90	176	150	28	27 580	200	75	10[b]
PVC	Unplasticised	1.40	–	80	59	58 608	5	–	–
	(vinyl chloride-co-vinyl acetate)	1.35	–	70	48	48 265	5	–	–
	Plasticised	1.31	–	Flexible	19	18 617	350	–	–
	Chlorinated	1.52	–	105	59	58 608	–	–	–
PS	General	1.04	–	100	42	42 060	2.0	–	19
	Medium impact	1.04	–	100	31	31 028	25	–	64
	High impact	1.05	–	100	18	17 927	60	–	110
	SAN	1.08	–	110	69	68 950	2.5	–	27
	ABS, medium impact	1.04	–	105	40	39 991	25	–	270
	ABS, high impact	1.04	–	106	37	37 233	25	–	400
PET	Fibre	1.38	–	–	690	689 500	15–50	–	–
	Film	1.38	–	–	170	172 375	70	–	–
	Amorphous moulding	1.30–1.34	–	–	55	55 160	250	–	53
	Crystalline moulding	1.32–1.38	–	–	76	75 845	250	–	43
Nylon	6,6	1.14	264	75[c]	79	79 293	80–100	–	27–53
	6	1.13	215	60[c]	76	75 845	100–200	–	32–53

[a] T_m: crystalline melting point.
[b] Impact strength (kJ m^{-2}) at −20 °C (no break was observed at 23 °C) (data adapted from Saunders, K. J. (1998). *Organic Polymer Chemistry*, 2nd edn, Chapman & Hall, London).
[c] Heat distortion temperature at 1.86 MPa (1.82× 10^6 N m^{-2}) (°C).

As the name suggests, LDPE has a low density (see Table 3.2), which is caused by a low degree of crystallinity. As we have already seen, this is related to the high level of side branching with both short and long chain branches. LDPE is used widely for its optical clarity and film forming properties and constitutes around 15 % of the world polyolefins market.

Linear low density polyethylene (LDPE) and very low density polyethylene (VLDPE) are also commercially available. LLDPE is prepared *via* the solution or gas phase copolymerisation of ethylene and an α-olefin (1-alkene), such as 1-butene, 1-hexene or 1-octene in small quantities (*ca.* 5–12 % w/w for LLDPE). The presence of the α-olefin leads to the presence of only short (but not long) alkyl branches. LLDPE also makes up around 15 % of the world polyolefins market. VLDPE is now commercially available and is prepared in the gas phase with higher levels of the α-olefin. VLDPE is more flexible than other forms of PE and is widely used for thin films, tubing and squeeze bottles.

High density polyethylene (HDPE) constitutes around 27 % of the world polyolefins market. It is usually produced by one of four methods. Three of these, *i.e.* the Ziegler, Phillips and Standard Oil processes (Figure 3.2), involve solution or slurry processes while the Union Carbide processes are conducted in the gas phase. Commercial Ziegler PE processes involve mild conditions in the presence of Ziegler–Natta catalysts to produce a polymer with significantly less branching, and hence a higher density, than corresponding high pressure processes. Typically, chains containing 5–7 ethyl groups *per* 1000 carbon atoms are produced (Table 3.2).

The Phillips processes are usually intermediate between the Ziegler and high-pressure processes and produce a polymer containing *ca.* 5 % chromium oxides (*e.g.* chromium trioxide) and chains containing up to three methyl groups *per* 1000 carbon atoms, but no ethyl or butyl residues. It is not known whether the Standard Oil processes proceed *via* a similar pathway to the Phillips, but the products from both do resemble each other, *i.e.* the PE produced is almost completely linear. The Union Carbide processes are somewhat different in that they take place in the gas phase within a fluidised-bed reactor and involve organochromium compounds. The process, which is essentially simpler and more energy efficient, yields the polymer in the form of a granulated product.

Two further forms of PE are high molecular weight high density polyethylene (HMW-HDPE) and ultra-high molecular weight high density polyethylene (UHMW-HDPE). These are also manufactured commercially where extremely good mechanical properties (*e.g.* high impact strength, resistance to environmental stress cracking and abrasion) are required. HMW-HDPE is produced using similar technology to the conventional HDPE, while UHMW-HDPE is prepared using a modified Ziegler process, in which the final product does not undergo granulation in order to avoid chain scission and molecular weight reduction.

The commodity PE is then processed into different product by the following principal techniques (ranked according to increasing PE molecular weight):

(i) Injection moulding is used for PE with high melt index (low viscosity) and narrow MWD. The viscous resin is squirted, by a means of a plunger, out of a heated cylinder into a water-chilled mould, where it is cooled before removal. Processing is performed from around 170 to 300 °C, depending on the material, but all begin to decompose above 310 °C (and very short processing times are required beyond this temperature). PE does undergo some degree of shrinkage during injection moulding (LDPE 1.5–3 % and HDPE 2–4 %). Injection moulding is used to produce a wide variety of domestic and industrial articles, which are too numerous to list.

(ii) Extrusion, in which melted PE is extruded at 160–240 °C to produce pipes using single- or twin-screw extruders and with mould temperatures of up to 300 °C to produce film, sheeting and monofilaments.

(iii) Compression moulding, whereby powder or granules of semi-finished product are put directly between heated tool faces, which are brought together under pressure and the material is shaped.

Key Facts
- LDPE is produced by solution *via* gas phase copolymerisation with an α-olefin.
- LLDPE is produced by solution, slurry or high pressure gas phase polymerisation.
- Domestic and industrial articles are produced by injection or compression moulding or extrusion at temperatures up to 310 °C.

In 1991, Western Europe consumed *ca.* 2.8 million tonnes of PE in the key areas of blow mouldings (39.3 %), injection mouldings (26.7 %), films (19.7 %) and pipes (14.3 %)[2]. The widespread use of PE is due to the favourable properties (*e.g.* relatively low density, good toughness and elongation at break, good electrical and dielectric characteristics, low water absorption and low water vapour permeability and the high resistance to chemical attack and environmental stress cracking, *etc.*) exhibited by the polymer. In 2000, growth in the world market for PE slowed (to around 3.9 %, compared with 8 % in 1999), but world demand for all types of PE surpassed 50 million tonnes for the first time[3]. It has been suggested[3] that this was due to the market growth being strongest in the less developed regions of the world in 2000 and weakest in the developed world (whereas the reverse was true in 1999, with North America, Western Europe and Japan growing at around 6.5 % compared with a more usual rate of 4.5 %).

3.2.2 Polypropylene (PP)

Polypropylene is produced from the propylene (more correctly termed propene) monomer as shown in Figure 3.1 by a similar chain growth route to that for polyethylene. It has a similar structure to PE except that every alternate carbon atom has a methyl ($-CH_3$) bonded to it instead of $-H$ (see Table 3.1). While PP homopolymer is generally the most common form of PP (making up *ca.* 70 % of the PP sold in Western Europe), the PP oligomers are also copolymerised with ethylene in a variety of compositions to form:

● block copolymers (around 25 % of the PP market in Western Europe);
● random block copolymers, impact PP (containing a blend of PP and PE/PP);
● polyolefinic thermoplastic elastomers.

The three main types of PP (highly isotactic homopolymers, block copolymers and random copolymers) may have similar rheological properties whilst displaying a wide range of physical and mechanical properties (see Table 3.4). This makes PP very versatile and this feature, coupled with its relatively low price, makes PP an increasingly attractive and important commercial thermoplastic. Ironically, the low price of virgin PP does reduce its potential for recycling, as discussed later, in Chapters 5 and 6.

Table 3.4 Selected properties of polypropylene homo- and copolymers.

Property	PP homopolymers	PP/PE copolymers	Block	PP/PE copolymers	Random
		Medium toughness	High toughness	Stiff	Soft
Young's modulus (MPa)	1400	1250	700	800	500
Impact strength $(kJ\,m^{-2})^a$ $-20\,°C$	10	90	2	13	50
Ball indentation hardness (MPa)	80	57	31	50	32
Melting point (°C)	162	162	162	146	130
Transparency[b] (%)	35	5	2	52	65
MFR 230/2.16[c] $(cm^3/10\,min)$	7	6	3.5	2.5	7

[a]No break was observed at 20 °C.
[b]Using 1 mm injection moulded disks.
[c]Melt flow rate.
Source: data adapted with permission from J. Kabovoc (ed.) 'Recycling of Polymers'. In *Macromolecular Symposia*, 135 (1998). Copyright (1998) Wiley-VCH, Weinheim.

PP is typically produced on an industrial scale by the same manufacturers that produce PE, with the use of Zieger–Natta catalysts or *via* either gas or liquid phase processes. The high pressure, free radical process has a low reactivity towards PP due to resonance stabilisation in the propagating radical, while the Phillips and Standard Oil processes (mentioned in the preceding PE sections) both produce low product yields. The Ziegler–Natta process does produce some atactic polymer in addition to the desired isotactic product (the latter

Key Facts
● In 1991, Western Europe consumed *ca.* 2.8 billion tonnes of PE.
● PP is widely used as both the homopolymer and copolymerised with ethylene.

having superior mechanical properties) and this is generally removed by extraction into heptane. Modern gas phase processes yield much lower quantities of atactic polymer. The properties and solubility of PP (Tables 3.3 and 3.4) are similar to those of HDPE, but with a higher softening point and a greater retention of some properties (e.g. tensile strength and stiffness) at elevated temperatures. PP is more liable to be oxidised in air (at elevated temperatures) than PE, due to the presence of a tertiary hydrogen atom on alternate carbons, but may be protected by the addition of anti-oxidants (see Chapter 2). For instance, well-stabilised PP can withstand more than 4000 h (6 months') of exposure in air at 150 °C or provide over 10 years of service at a sustained temperature of 100 °C [4]. The resistance of PP to weathering can also be improved so that the polymer may survive for over 10 years or more without suffering significant degradation.

Key Facts
● PP has similar properties to HDPE, but is more stable.
● In 1996, the world production of PP was ca. 16 million tonnes.

PP experienced a substantial growth in market sales between 1980 and 1992 (ca. 10.2 % p.a., a doubling of growth every 7 years)[4]. In 1992, the world output of PP was ca. 13 million tonnes (this rose to 16 million tonnes in 1996)[4], making PP the third most important commodity polymer, behind PE and PVC. Of the 1992 total, around 4.7 million tonnes of PP were produced in Western Europe (with 17 major manufacturers competing for market share) while the market consumption was ca. 4.2 million tonnes. In 1999, this figure rose to 6.7 million tonnes, an increase of almost 40 %. PP is used in the key areas of injection mouldings (47 %), fibres (18 %), extruded flat tape (11 %), oriented PP (10 %), cast film (6 %), thermoformable sheet (2 %) and other miscellaneous uses, such as blow mouldings (5 %)[4]. Critically, the price of PP fell dramatically in 1993, making recycling of PP less attractive and leading to rationalisation and a reduction in the number of Western European producers. In February 2001, there were ten producers in Western Europe, with Basell by far the largest ahead of Borealis[3]. In 2000, growth in the world market for PP slowed, yet world demand for PP still reached 30 million tonnes. Despite the economic slowdown, the growth of the PP market is still forecast to outstrip the growth of other commodity polymers, such as PE, PVC and PS.

3.2.3 Polystyrene (PS)

Polystyrene is sold in three main forms: crystalline or general purpose polymer, high impact polymer and expanded polymer. It is the combination of low cost, ease of fabrication (due to its good flow properties), transparency/colour fastness and high surface gloss that has led to the wide commercial success of PS. However, whilst the polymer possesses great rigidity, this is accompanied by low impact strength, although this drawback may be overcome as discussed below in the section on formation of 'high impact PS'. The commercial preparation of the monomer, styrene, is shown in Figure 3.4.

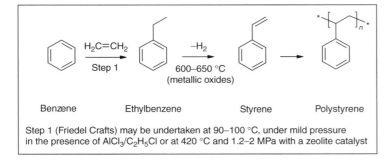

Figure 3.4 Preparation of styrene monomer and polymerisation to form PS

Styrene may be polymerised by a variety of methods, including bulk processes which produce a relatively polydisperse polymer. Other methods include solution, suspension or emulsion processes, which are used to a limited degree to produce PS latex for water-based surface coatings. Free radical polymerisation routes produce predominantly non-crystalline, syndiotactic PS, while Ziegler–Natta catalysts and *n*-butyllithium produce a highly crystalline form that is difficult to process as a result. Commercial PS has a 'head-to-tail' form (see Chapter 2 for an explanation) with an average molecular weight typically in the range 50 000–200 000 and is a hard and relatively brittle polymer (Table 3.3).

Expanded polystyrene (EPS), formed from 'expandable beads' of pre-polymerised PS, is the ubiquitous thermal insulating material with a thermal conductivity comparable with cork and glass wool. The raw material for EPS is produced by suspension polymerisation in a stirred vessel to produce beads. This is followed by impregnation with a blowing agent, along with additives to influence the foam properties and organic (*e.g.* poly(vinyl pyrrolidinone), PVP) or inorganic (*e.g.* tricalcium phosphate) stabilisers. The formation of the foam from the raw material involves three stages:

(i) pre-foaming, whereby raw material beads are expanded at *ca.* 100 °C to around 60 times their original volume;
(ii) interim storage whereby the beads cool and are stabilised;
(iii) final foaming in which pre-foamed, stored particles are treated with steam to expand the beads and to form a homogeneous foam as the blowing agent escapes from the beads, prior to final processing.

It is possible to modify the brittleness of PS by the introduction of *e.g.* poly(*cis*-buta-1,4-diene) (forming 'high impact PS', HIPS), although optical clarity and tensile strength do suffer as a result. PS is also copolymerised with *ca.* 20–30 % acrylonitrile *via* a bulk or suspension route to produce a random, amorphous copolymer, poly(styrene-*co*-acrylonitrile), SAN. The resulting material has a higher softening point and higher impact strength than PS homopolymer but, while still transparent, it does carry a faint yellow pigmentation making it unsuitable for some household/food packaging applications. Finally, ABS, the product of the blending or graft copolymerisation of acrylonitrile, butadiene and styrene, has found wide commercial importance, particularly in the injection moulding or extrusion of polymers with high toughness and rigidity (Table 3.3). For example, SAN finds application in appliance knobs, refrigerator compartments and syringes, *etc.* while ABS is now widely used for vehicle fascia panels and radiator grilles, and household appliances, such as telephones and pipes/pipe fittings.

The PS market is a large one: the capacity for the polymer was 8.6 million tonnes in 1991 with manufacturers based in North America (having a 32.6 % share of the total capacity), East Asia (31.4 %), Western Europe (26.7 %), Latin America (7 %) and Africa (2.3 %). From 1986 to 1990, the consumption of PS in Western Europe grew steadily from 1 million tonnes to 1.8 million tonnes, but fell slightly to 1.76 million tonnes thereafter, due to the general downturn in the economy[5]. In 2001, the world demand for EPS, for example, was at *ca.* 1.6 million tonnes.

3.2.4 Poly(vinyl chloride) (PVC)

Poly(vinyl chloride) (also known as V, PVC or vinyl) is formed by polymerisation of vinyl chloride monomer (VCM) in a process similar to that for PE, PP and PS. It tends to polymerise in the 'head-to-tail' structure depicted in Figure 3.5 and, in its virgin form, is a colourless

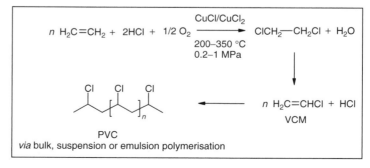

Figure 3.5 Preparation of vinyl chloride monomer (VCM) and polymerisation to form PVC

Key Facts
● PS is easy to fabricate with colour fastness and high gloss finish but has low impact strength.
● Expanded PS has low thermal conductivity equivalent to cork or glass wool.
● High impact PS is produced by the addition of dienes.
● ABS is made from acrylonitrile, butadiene and styrene and is widely used in vehicles.
● In 1991, 8.6 million tonnes of PS were consumed world wide.

rigid polymer with a range of useful physical properties (see Table 3.3). Commercial production usually takes place *via* suspension polymerisation to produce suspension or S-PVC in which grain shape and size guarantee good flow properties. A porous grain is employed to ensure high plasticiser absorption and hence plasticised PVC, while a compact grain produces rigid-PVC. Bulk and emulsion methods are used less frequently, while solution methods are seldom used. Emulsion PVC (E-PVC) is produced as a fine powder containing residual emulsifier, making it the polymerisation route of choice for paste processing to produce textile coatings (in the form of 'leathercloth'). Mass PVC (or M-PVC) is produced in a more dense, powdered form than S-PVC in higher purity, making it particularly attractive in transparent moulding applications. PVC is also copolymerised with suitable comonomers to improve the thermoforming properties and impact resistance or heat resistance of the homopolymer. In this way poly(vinyl chloride-*co*-vinyl acetate)s (containing 5–20 % vinyl acetate) are used for film manufacture. Poly(vinyl chloride-*co*-acrylic ester)s (containing 6–12 % of the acrylic ester) are employed in high impact, weather resistant window frames.

As VCM is a human carcinogen, it is important that the PVC product contains as little residual monomer as possible. During the latter part of the 1970s, the production process was significantly improved to address this problem, so that now polymer is routinely produced with less than 5 ppm VCM content (or <1 ppm VCM for food packaging applications where the polymer may come into direct contact with food). Chemically, PVC is relatively inert, although exposure to either ultraviolet light or heat leads to degradation, leading to discolouration and changes in mechanical properties. Aspects of degradation are covered in greater detail in Chapter 2, but the reactions by which this process occurs in PVC are complex and not completely understood.

In 1999, consumption of PVC was running at more than 5.7 million tonnes in Western Europe[6] (and 5.2 million tonnes in 1995[7]) having risen steadily from a figure of around 2.35 million tonnes in 1970. This places it just behind PE and among the most important of the commodity thermoplastics. The production of PVC is largely made up of S-PVC (75 %), E-PVC (13 %), with the remaining 12 % being made up of M-PVC and its copolymers. Major uses of PVC are listed in Table 3.1.

Key Facts
● PVC is chemically inert but susceptible to UV light.
● In 1996, Western Europe consumed more than 5.2 million tonnes of it.
● PET has good chemical resistance and toughness, excellent transparency and gloss and high barrier properties to gas, but also a high T_g of 80 °C.

3.2.5 Poly(ethylene terephthalate) (PET or PETE)

The starting compounds for the commercial production of PET are ethylene for the production of ethylene glycol and *para*-xylene (*p*-xylene) for the production of terephthalic acid. As shown in Figures 3.1 and 3.6 there are two routes for the production of PET. In the first, *p*-xylene is oxidised to terephthalic acid (R = H) which is then reacted with ethylene glycol to produce *bis*(2-hydroxyethyl)terephthalate (BHET) by direct esterification. In the alternative route, *p*-xylene reacts with methanol to produce dimethyl terephthalate (R = CH$_3$). Ethylene glycol is then reacted in an ester exchange reaction with dimethyl terephthalate to produce BHET (see Figure 3.6). The step growth polymerisation of BHET gives amorphous PET, which is suitable for fibres and film. A second polymerisation step produces a partially crystalline polymer that can be moulded into bottles.

PET is a colourless, rigid crystalline polymer, stemming from the high degree of structural regularity that may be present, depending on the processing steps undertaken during preparation, such as extrusion and drawing. The potentially high crystallinity may have a profound effect on the final properties of the polymer (Table 3.3), particularly its solubility: highly crystalline forms are soluble at room temperature in proton donors or halogenated solvents. The relatively high T_g (80 °C) also makes conventional injection moulding problematic without the addition of nucleating agents (to promote crystallisation following cooling after moulding) and plasticisers (to reduce T_g). The properties that particularly characterise PET are very good chemical resistance, good toughness/shatter resistance, very high mechanical properties, excellent transparency and gloss, and high barrier properties (especially for oxygen and carbon dioxide). This combination of properties renders PET particularly attractive for food and beverage packaging and the polymer

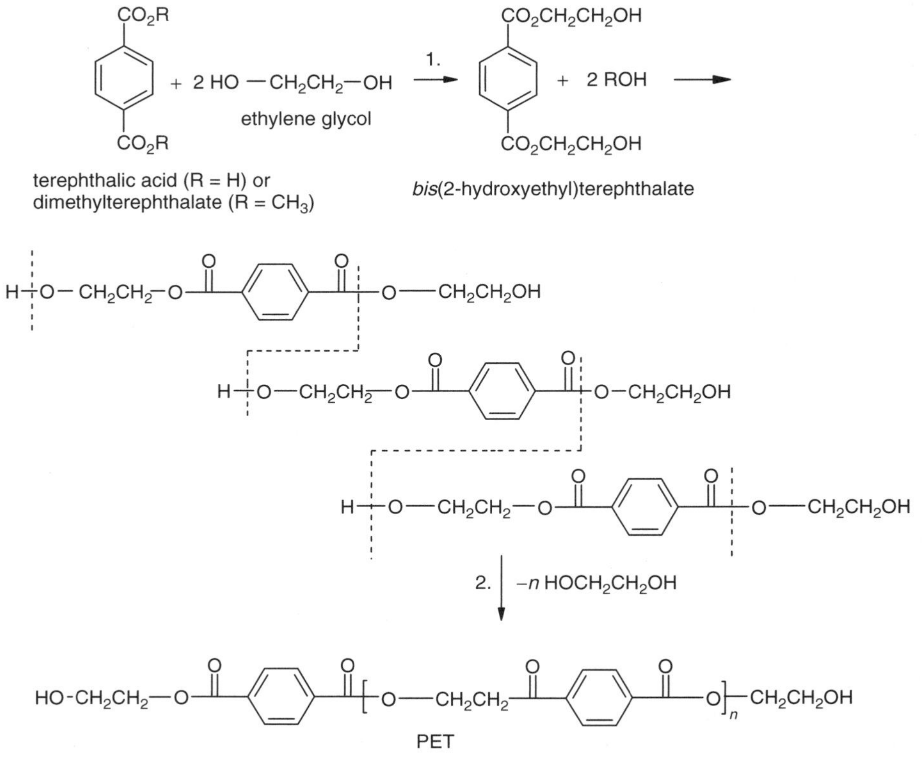

1. R = H: Excess ethylene glycol (1:1.5 molar), 250 °C, 0.4 MPa
   R = CH_3: Excess ethylene glycol (1:2.2 molar), 140–220C, atmospheric/manganese acetate

2. 290 °C, SbO_3, 0.1 kPa

Figure 3.6 Preparation of PET *via* ester interchange

Key Facts

- In 1995, the world production of PET was 16.5 million tonnes and 2.9 million tonnes of PET were used in bottles and food packaging.

is approved by the Food Drug Administration (FDA) and BGA (Federal Office of Health, Germany) for food contact. PET displays good resistance to photochemical degradation (see Chapter 2), although it does suffer some thermal degradation when heated above its crystalline melting point of 265 °C.

In 1995, the world consumption of PET amounted to *ca.* 16.5 million tonnes[7], with the Western European market consuming 2 million tonnes of this total. In 1999, the consumption in Europe grew to 2.9 million tonnes[6], representing an increase of 40 % from the volume in 1995. The principal uses of PET are in fibre applications (as fibrefill, filaments or staple fibres, *etc.*) but the polymer is also used in technical applications, such as audio and video cassettes (see Table 3.1). In 1995, 2.9 million tonnes were consumed worldwide in packaging applications across a wide range of areas including bottles for carbonated drinks, mineral water, edible oil, cosmetics, surfactants, films for thermoforming applications and packaging tape, *etc.*[8].

3.2.6 Other Polymer Types

The remaining commercial polymers are a diverse group comprising among others, thermoplastics such as poly(methyl methacrylate) (PMMA), acetals, polycarbonates, polyamides, and acrylics, *etc.*, and thermosets such as alkyds, amino resins, phenolics, epoxy resins, polyesters (other than PET) and polyurethanes (PUs)[9]. A detailed description of their processing routes is outside the scope of this book, but suffice it to say that they are also mainly derived from nonrenewable fossil sources and that their production involves numerous precursors and intermediates. For example, precursors for polyurethane are normally tolylene-2,4-diisocyanates (TDI) and 4,4′-methylene-*bis*(phenylisocyanate) (MDI) and polyols. The process routes for their production are considerably more complex[10] than for the tonnage thermoplastics described above.

3.3 GLOBAL CONSUMPTION OF POLYMERS

Paradoxically, while numerous polymers have been synthesised on a laboratory scale, the international polymer market is still largely based on the production and consumption of a relatively small number of polymeric materials. Among these, commodity thermoplastics, listed in Table 3.1, have the largest market share. They are used in a variety of primary applications and can be recycled into a number of different products.

The production and consumption of plastic materials have been growing steadily over the past few years. For instance, from 1995 to 1999 the production of five principal thermoplastic polymers (LDPE/HDPE, PP, PVC, PS and PET) grew by 15%[6,7]. In 1999, the 'big five' accounted for 79% of the total polymer consumption (thermoplastics and thermosets) in Western Europe and around 90% of thermoplastics[6]. In the same year, their consumption exceeded 28 million tonnes overall (see Figure 3.7). This is significantly larger than the production in the next largest sector, *i.e.* polyamides (at 1.3 million tonnes, less than half the market volume of PET) followed by ABS/SAN (646 ktonnes). The same trend was followed in the rest of the world, with the total market volume exceeding 100 million tonnes of thermoplastic materials[11].

Earth Image

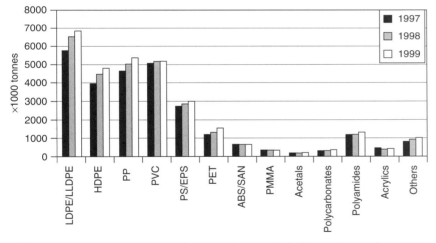

Figure 3.7 Production and consumption of thermoplastic materials in Western Europe from 1997 to 1999[6]. Reproduced by permission from APME (2001). An Analysis of Plastics Consumption and Recovery in Western Europe 1999. Copyright (2001), APME, Brussels

Key Facts
● In 1999, LDPE/HDPE, PP, PVC, PS and PET consumption exceeded 28 million tonnes in Western Europe, 79% of the total plastics market.
● In the same year, the total consumption of thermosets was only 8.1 million tonnes in Western Europe.

In contrast, the thermoset polymer market is comparatively small, with the 1999 consumption in Western Europe just exceeding 8.1 million tonnes[6]. The sector was largely represented by five different families of polymers (in decreasing order of consumption): amino resins, polyurethanes, phenolics, polyesters, alkyd resins and epoxy resins (see Figure 3.8). However, the market is still significant since some of the properties offered by higher performance thermoset polymers are generally unmatched by commodity thermoplastics. Consequently, some market sectors (*e.g.* the aerospace and microelectronics industries) rely heavily on these materials despite consuming relatively small tonnages. While technologies for recycling of thermoplastic materials are fairly well developed, thermoset polymers present altogether different (and generally more demanding) challenges. This is mainly due to the three-dimensional network formed during cure (see Chapter 2)

Key Facts 🐛━━━━⚿

● The primary use for plastics is packaging, followed by building, electronics and automotive industries.

● The total consumption of all plastics in Western Europe in 1999 was 84 kg *per* person.

and the resulting insolubility and high heat resistance. The recycling options for these and other polymeric materials will be discussed in Chapter 4.

In general, the plastics consumption figures for the three developed geographical regions shown in Table 3.5 exhibit broadly similar distributions, with packaging being a significant market in all regions. This remained so in the following years and in 1999, the share of the packaging market in Western Europe increased to 40 %. Another area that has seen an increased use of plastics is the automotive industry, whose share grew to 8 %. The total plastics consumption in Western Europe in 1999 was 33.6 million tonnes or 84 kg of plastics *per* person[6]. A breakdown of consumption by different European countries and by sectors is shown in Figures 3.9 and 3.10, respectively. Of that amount, 19.2 million tonnes were available for collection as waste. However, only 6 million tonnes, or 30 % of the total post-consumer waste were recycled[6]. The following section gives an overview of sources of polymer waste.

Table 3.5 Plastics consumption by sector in different regions in 1991.

Category	Region		
	USA (%)	Western Europe (%)	Japan (%)
Packaging	28	32	30
Building	22	22	10
Electronic/electrical	6	9	14
Automotive	3	4	9
Other transport	1	3	1
Other markets	38	30	38

Source: data adapted from Kirkwood, R.C. and Longley, A.J. (eds) (1995) *Clean Technology in the Environment*, Copyright (1995) Kluwer Academic Publishers.

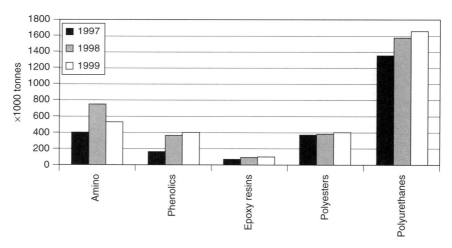

Figure 3.8 Production and consumption of thermoset materials in Western Europe from 1997 to 1999[6]. Reproduced with permission from APME (2001). An Analysis of Plastics Consumption and Recovery in Western Europe 1999. Copyright (2001), APME, Brussels

3.4 WASTE STREAM CATEGORIES

At the end of their useful life, waste polymers enter waste streams as either post-consumer waste or industrial scrap. Households and the distribution and industry sector are the sources of the former while the latter arises from processing, filling, assembling, installing and polymerisation. Much of the industrial waste is recycled within the process and the rest is usually sent for reprocessing by a third party. Consequently, little of this material

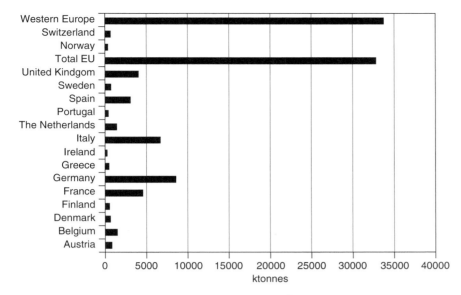

Figure 3.9 Plastics consumption by country in Western Europe[6] in 1999. Reproduced with permission from APME (2001). An Analysis of Plastics Consumption and Recovery in Western Europe 1999. Copyright (2001), APME, Brussels

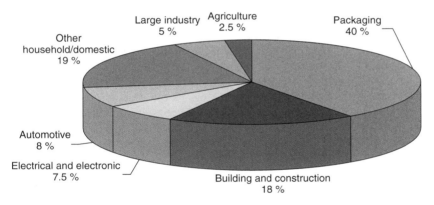

Figure 3.10 Consumption of plastics by sector[6] in Western Europe in 1999. Reproduced with permission from APME (2001). An Analysis of Plastics Consumption and Recovery in Western Europe 1999. Copyright (2001), APME, Brussels.

[Agriculture:	0.85 million tonnes
Automotive:	2.25 million tonnes
Building and construction:	6.17 million tonnes
Electrical and electronic:	2.54 million tonnes
Packaging:	13.5 million tonnes
Total consumption	33.6 million tonnes]

is discarded as waste. The majority of post-consumer plastics waste on the other hand reaches the environment and hence the emphasis in polymer waste management is on this type of waste stream.

Post-consumer plastics waste (particularly in a commingled form) may arise from a host of products or applications each with differing life cycles. Stein[12] gives an interesting explanation of why polymers become mixed. He states that, 'without intervention, polymeric articles will tend to become mixed, increasing disorder and entropy. Demixing to decrease entropy requires an expenditure of energy.' He points out that a decision must be made whether the energy expended in reducing the entropy and raising the polymer article's free

Key Facts 🐭———✂

● Most industrial waste is recycled, whereas most consumer waste is not.

● Separating mixed polymer waste is difficult; recycling may consume more energy than can be recovered from combustion of polymers.

energy state compares favourably with the energy recouped from, for instance, combustion of the article to generate heat or power. Therefore, it is very difficult to separate the waste once mixed and to achieve a consistent quality of waste stream. These are some of the major constraints that influence recycling. The use of polymers in different market sectors further complicates the identification and separation of waste streams because their use in one sector does not guarantee that they will appear as waste in the same sector. Hence, it is more appropriate to devise the following six broad categories of waste streams in which polymers may be found:

  (i)   municipal solid waste;
  (ii) automotive waste;
  (iii) construction waste (including demolition and civil works);
  (iv) distribution and large industrial waste;
  (v)   agricultural waste;
  (vi) electrical and electronic waste.

3.4.1 Municipal Solid Waste (MSW)

Municipal solid waste covers the largest volume of plastics waste

The definition of MSW is broad and includes household waste (of a wide variety from coat hangers to food packaging and large domestic items) and waste from commercial activities and retailers (e.g. wholesalers and supermarkets). This category currently covers the largest volume of plastics waste (accounting for around 12.8 million tonnes in Western Europe in 1999, around 67 % of the total produced from all categories)[6]. However, polymers make up a weight fraction of only 7–8 % in the total MSW stream (and only 1 % in the total waste). Packaging materials make up the largest contribution to polymer waste in MSW (ca. 65 %)[13]. There are six major plastics in the domestic solid waste stream (in decreasing order by quantity): LDPE/LLDPE, PP, HDPE, EPS/PS, PVC and PET. In addition to these, there are another 100 categories that could be sorted from household waste and this represents one of the main problems for the packaging industry and recycling.

However, it should be noted that abandoning the use of plastic packaging would not necessarily result in a reduced impact on the environment. For example, it has been stated[14] that the use of alternative packaging would result in a 404 % increase in the weight of waste, a 201 % increase in energy consumption in making the alternatives and a 212 % increase in cost. The weight of plastic packaging is also being constantly reduced through improved design and manufacturing practices. This has led to a decrease of packaging weight by 28 % in the past 10 years, so that now plastic packaging accounts for only 17 % of all packaging by weight. In theory, this should lead to a reduction in the amount of waste produced; however, in reality this does not happen because of the ever-increasing consumption which pushes up the waste volumes. For example, between 1998 and 1999, consumption of polymers rose by 5.4 % in Western Europe alone[6].

3.4.2 Automotive Waste

Plastics have increasingly been used to replace metals for many automotive components because of their greater processability, lighter weight (leading to lower fuel consumption) and corrosion resistance, etc. Although the percentage by weight of polymers in the average European car has risen from 2 % to 12 % in recent years, in contrast with the packaging applications, the number of polymer types used in the automotive industry is relatively small. Nevertheless, around 2.3 million tonnes of polymers were used in this application in Western Europe in 1999, around 8 % of the total produced from all categories[6]. Polymers find use in both bulk and glass fibre-reinforced composite form (e.g. PP or ABS bumpers, HDPE fuel tanks,

Many automotive components such as car seats are now made from polymers

> **Key Facts** 🔑
> ● In 1999, Western Europe produced 12.8 million tonnes of MSW, of which 7–8 % were plastics, mainly packaging.
> ● 2.3 million tonnes of plastics were used in the automotive industry.

elastomeric tyres, glass reinforced PP and nylon, PU foam seating, PVC in under-floor protection, *etc.*). New specifications for the materials develop opportunities for the use of polymers; however, some technical difficulties, such as problems with surface finish on polymer body panels, still make recycling difficult and limit the re-use of polymers[15]. Assisted by the proposed EOL Vehicle Directive (see Chapter 1), interest in recycling automotive components is growing, along with the design of items employing single grade thermoplastics for ease of recycling.

House Under Construction. Polymers are used extensively in the construction of modern houses

3.4.3 Construction Waste (Including Demolition and Civil Works)

The construction industry makes extensive use of PVC in a variety of applications (*e.g.* piping, flooring, window frames and electrical cables), while thermosetting resins (phenolics or amino resins) are used, together with acrylics or PP, for kitchen or bathroom fittings. Inevitably, significant quantities of construction waste become commingled with MSW. This category accounted for around 6.2 million tonnes in Western Europe in 1999, around 18 % of the total produced from all categories[6].

Packaging Bales. Large-scale industry tends to generate clean, largely uncontaminated waste

3.4.4 Waste from Distribution and Large Industry

PE sheet is used almost exclusively for all secondary packaging in the distribution industry. The nature of the business generates clean, largely uncontaminated waste, which can be easily used to form recyclate of good quality. In large-scale industry, the plastic waste involved often consists of a single polymer type (and in some cases also of a single grade) or at least a well-defined blend of components.

3.4.5 Agricultural Waste

Around 848 000 tonnes of polymers were used in agricultural applications in Western Europe in 1999, or about 2.5 % of the total produced from all categories[6]. The plastics waste is predominantly LDPE sheeting (used primarily for two applications, *i.e.* mulching or crop/fodder protection, *etc.*). Matthews[13] notes that, while the waste is relatively easy to collect and bale, the presence of large quantities of soil/plant debris and contaminants may compromise quality significantly.

> **Key Facts** ✄
> ● 6.2 million tonnes of plastics were consumed in construction in 1999 in Western Europe, plus 0.8 million tonnes in agriculture and 2.5 million tonnes in the electrical/electronic industry.

3.4.6 Electrical and Electronic Waste

With around 2.5 million tonnes of plastics used in Western Europe in 1999, or 8 % of the total in all sectors, the electrical and electronics industry is a significant user of plastics. The main use of polymers is in televisions, video cassette recorders, microwave cookers, personal computers and telecommunications. The major polymers used are poly(acrylonitrile-*co*-butadiene-*co*-styrene) (ABS) (40 %), polycarbonate and modified PP (18 % each). For example, in 1995 telecommunications consumed about 30 000 tonnes of plastics, of which it was estimated that 85 % were used for telephone handsets, mainly produced in ABS[14]. Given the rate of expansion of this industry, particularly the exponential growth of the mobile telephones sector, this amount looks set to rise, despite the reduction in the amount of plastics used *per* item. There are also significant amounts of thermoset polymers present in almost every item of electronic equipment, *e.g.* support

PCBs processing. The production of printed circuit boards requires very clean conditions

Crop Harvest. The agricultural industry tends to use polymers in mulching or crop/fodder protection purposes

material in printed circuit boards[14]. These are required to exhibit relatively high performance (*e.g.* to maintain structural rigidity at solder processing temperatures) and to display low dielectric loss properties and so polyimides, polycyanurates and epoxy resins tend to be favoured.

This brief overview of waste categories highlights the diversity of waste sources and the resulting difficulties in recovering, separating and recycling polymers. In the next sections, we review the amounts of waste generated around the world and outline the recovery and recycling practices in different countries to show how some of the recycling difficulties can be overcome by suitable national and international policies.

3.5 RECOVERY AND RECYCLING OF POLYMER WASTES IN DIFFERENT COUNTRIES

The problem of waste management is a global one and different countries have adopted different strategies to minimise its impact on the environment. In this section we will examine case studies from across the world and, while the coverage is not exhaustive, it does give an indication of the variation that is observed within different geographical areas.

3.5.1 Western Europe

Table 3.6 shows the total post-consumer plastics waste generated within Western Europe in 1995[10], separated into the categories discussed earlier. An indication of the population of the country (based on estimates in 2000) and waste generation *per capita* are also included for comparison. On average, each person in Western Europe generated approximately 24 kg of plastic waste that year, with Portugal and Greece being the most 'wasteful' countries at 43 and 39 kg, respectively, and the Netherlands being the most 'sustainable' at 19 kg *per capita*. In 1999, the figure for total waste rose from 16 million tonnes to 19 million tonnes[6]. This compares with the consumption of plastic in 1999 of 33.6 million tonnes, which means that only a half of that reached the waste stream that year. This is the usual

> **Key Facts**
> ● In 1999, Western Europe produced some 33.6 million tonnes of plastics and generated 19 million tonnes of waste plastics.

Table 3.6 Total post-consumer plastics waste generated in Western Europe in 1995 (data adapted with permission from APME (1997). Plastics Consumption and Recovery in Western Europe in 1995. Copyright (1997) APME, Brussels).

Country	Population[a] (million)	Total waste 10^3 tonnes year^{-1} (tonnes *per year* and *capita*)	Waste stream category					
			MSW	Distribution/ large industry	Automotive	Agriculture	Construction/ demolition/ civil works	Electrical/ electronic
Austria[b]	8.15	275 (30)	157	60	18	7	17	16
Belgium/Luxembourg	10.62	526 (20)	256	128	32	10	59	41
Denmark	5.24	219 (24)	97	75	12	6	14	15
Finland[b]	5.16	242 (21)	147	51	11	6	12	15
France	59.13	2950 (20)	2009	439	172	46	153	131
Germany	82.08	3131 (26)	1771	706	194	53	224	183
Greece	10.75	279 (39)	199	52	8	7	6	7
Ireland	3.65	162 (23)	96	42	7	3	8	6
Italy	56.69	2411 (24)	1615	397	130	65	95	109
Netherlands	15.88	834 (19)	494	204	39	11	46	40
Norway[b]	4.45	141 (32)	75	40	10	3	7	6
Portugal	9.90	228 (43)	106	87	10	5	6	14
Spain	39.21	1740 (23)	1334	218	54	28	39	67
Sweden[b]	8.94	423 (21)	282	79	18	7	16	21
Switzerland[b]	7.29	337 (22)	222	59	22	4	17	13
United Kingdom	59.25	2158 (27)	1279	446	151	32	122	128
		16056 (24)	10139	3083	888	293	841	812

[a]Population statistics (2000 estimates) derived from http://encarta.msn.com/.
[b]Not a member of the EC at the time of analysis.

trend, because there is a time lag between production and end of life, depending on the particular polymer and its application.

The generation of such large quantities of waste material has the potential to create a significant problem in the management of plastics waste. As discussed in Chapter 1, the European Commission (EC) has already recognised this problem and, under the umbrella of producer responsibility legislation, it has set ambitious targets for waste management. The examples include the EC Directive on Packaging and Packaging Waste[16] and the Waste Electrical and Electronic Equipment (WEEE) Directive[17]. The former aims to reduce the amount of packaging waste going to landfill sites by setting targets for recycling and recovery. The WEEE Directive sets out measures that aim, firstly, at the prevention of WEEE, secondly at the re-use, recycling and other forms of recovery of such waste, and thirdly at minimising the risks and impacts to the environment associated with its treatment and disposal.

In 1994 the total figure for plastics recovered from all sources in Western Europe was only 19.7 %[18]. The amount recovered each year has increased since then, though 1999 figures indicate that total recovered plastics represents just over 30 % of the waste stream[6]. Figure 3.11 shows the amounts of plastic waste recovered and recycled in Western Europe in 1999[6]. Evidently, the majority of post-consumer waste was recycled in the waste-to-energy schemes, followed by mechanical recycling and much lower rates by feedstock recycling. It has been estimated, based on markets for certain recycled materials and the availability of appropriate waste streams, that mechanical recycling in Europe has the potential to grow to 2.7 million tonnes by 2006[19].

Key Facts
● In 1999, most plastic waste was recycled via energy recovery schemes.
● The 'green dot' scheme in Austria, Belgium and Germany signifies that a product is recyclable and that the producer has paid a fee for its recycling costs.

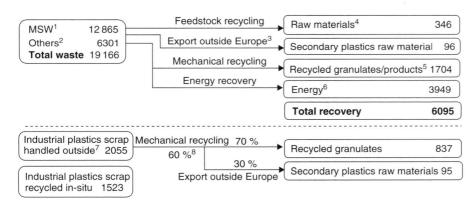

Figure 3.11 Recycling of plastic waste in Western Europe in 1999[6] (in ktonnes)

[Notes:

1 Households and assimilated
2 Distribution and industry
3 Mainly Asia and Central Europe
4 Paraffin methanols
5 85 % granules, 15 % plastic products
6 Of which 150/200 000 tonnes energy recovered in cement kilns
7 Processing, filling, assembling, installing, polymerisation
8 40 % is not recycled]. Reproduced with permission from APME (2001). An Analysis of Plastics Consumption and Recovery in Western Europe 1999. Copyright (2001), APME, Brussels.

As discussed below, several European countries have already employed collection schemes with varying degrees of success to recover common bulk thermoplastics (e.g. HDPE, PET and PVC). Many of these schemes have focused around the recovery of packaging waste following the implementation of the EC Directive on Packaging and Packaging Waste Directive (Directive 94/62/EC) and the following sections mainly examine schemes that enable collection and recycling of packaging. A summary of the amount of packaging waste and its recycling rates for the EU countries in 1997 and 1998 is given in Table 3.7[20].

Table 3.7 The amount of packaging waste and recycling rates in Western European countries from 1997 to 1998[20]. Reproduced with permission from PIRA (2001). Evaluation of costs and benefits for the achievement of re-use and recycling targets for the different packaging materials in the frame of the Packaging and Packaging Waste Directive 94/62/EC, Proposed Draft Final Report. RDC, Environment & PIRA International, May, p. 341. Copyright RDC, Environment & PIRA International (2001).

	1997		1998	
	Waste (10^3 tonnes)	Recycled (10^3 tonnes)	Waste (10^3 tonnes)	Recycled (10^3 tonnes)
Austria	180	36 (20 %)	190	41 (22 %)
Belgium	208	53 (25 %)	N/a	N/a
Denmark	183	11 (6 %)	172	N/a
Finland	90	9 (10 %)	90	9 (10 %)
France	1 571	102 (6 %)	1 628	131 (8 %)
Germany	1 502	675 (45 %)	N/a	600
Italy	1 777	164 (9)	1 800	192 (11 %)
Luxembourg	7	N/a	9	N/a
The Netherlands	611	76 (12 %)	491	49 (10 %)
Spain	1 215	65 (5 %)	N/a	N/a
Sweden	1 501	21 (14 %)	140	N/a
United Kingdom	13 561	100 (7 %)l	13 161	115 (9 %)

Obviously, most countries, except Germany, were still far from the targets imposed by the Packaging Directive (see Chapter 1). It is anticipated that many new schemes will evolve in the near future as a consequence of the WEEE, End of Life Vehicles and other imminent EC Directives.

Austria

Austria and Germany have adopted similar approaches to the implementation and regulation of plastics recycling[21] and the Government holds producers, distributors and importers responsible for the packaging materials used. Under the current scheme, these organisations are obliged to find methods to re-use the packaging materials or implement an open system of gathering and recovery. All but the smaller producers and distributors (*i.e.* those introducing less than stated quantities of materials, *e.g.* 100 kg of plastics, into the market on an annual basis) are included in the 'green point' ('grüne punkt') scheme. In this scheme, companies pay a licence fee for being able to display the green dot or 'punkt' on their packaging. The green dot signifies that the manufacturer or distributor has paid a fee to the appropriate agency or organisation and that the packaging has the potential to be recycled. Energy recovery is also included in the scheme.

Belgium

Regional authorities have the responsibility (and legislate) for waste management in Belgium, although a degree of harmonisation over the regulations (in the form of a federal law concerning environmental protection and public safety) has been felt necessary to avoid discrepancies between economic operators and consumers in different regions[21]. Belgium has adopted a similar 'green point' system to the Austrian and German system, so that responsibility for packaging wastes falls on users and importers. For instance, organisations that introduce more than 10 tonnes of packaging annually into the market are expected to prepare and implement a 3-year prevention programme. Alternatively, industrial packaging users have the option to return used packaging materials to one of a number of organisations: original suppliers, the municipal body, or a national organisation, set up for this purpose. Finally, the industrial packaging users may choose to provide their own recycling or utilisation systems.

Key Facts
● Responsibility for recycling packaging in most European companies falls to users and importers.
● Denmark currently has over 40 incinerators for recycling by energy recovery.

The Danes adopted the 'Environmental Protection Act' in 1993, a piece of legislation that requires producers and importers to increase both lifetime cycle and recycling of their products and to ensure that their disposal does not involve any damage to the environment[21]. Similarly, it is expected that users and customers will also promote recycling. Plastics do feature in this legislation: the use of PET re-fillable bottles is obligatory in Denmark for bottling beer and all carbonated beverages. The use of PVC in packaging has also markedly decreased in response to pressure applied to Danish industry. Denmark already achieves the maximum level (100 %) of energy recovery by combustion (and has at least 40 incinerators for this purpose)[20,21].

France

The packaging ordinance set in France required that 75 % of all household packaging waste be recovered by 2002[22]. In response to this, French product manufacturers and retailers set up Eco-Emballages SA, an organisation which issues licences to companies who in turn obtain a trademark agreement. This agreement exempts participating companies from their legal obligation to recover packaging. These companies are also eligible to display the Point Eco-Emballages trademark, similar to the original system German 'green dot' (see below). In return, Eco-Emballages organises recovery and recycling of packaging by appropriate waste management companies and also offers local authorities the possibility of developing a collecting and sorting system themselves, both of which are paid for by the licences. Where the French system differs from the German system is that the former does not discriminate between different forms of packaging waste 'utilisation' (*i.e.* it considers re-filling, recycling and energy recovery in the same way). Several other initiatives have also been set up for other segments of the waste stream such as glass, wood and pharmaceuticals.

Germany

The German Packaging Ordinance passed on 12 June 1991 laid down even more stringent requirements than the aforementioned EC Directive on Packaging Waste as it requested that 64 % of the waste separately collected and sorted should be mechanically recycled[23]. However, Brandrup[23] acknowledged that the quality (*i.e.* complexity) of the plastic household waste made this difficult to achieve by purely mechanical means.

The Duales System Deutschland (DSD) aims to provide a means for achieving the targets set by the 1991 ordinance. The system involves the kerb-side collection of segregated waste from German households. (The waste is left on the kerb-side in front of the house, hence the name.) Waste products such as plastic packaging, aluminium cans, paper cartons and glass bottles are collected[22]. The 'green dot' waste must be recycled and under current rules cannot be incinerated.

The fees collected enable the kerb-side collection schemes to be operated and the companies are able to meet the recycling targets set by the 1991 ordinance without themselves becoming involved in a whole recycling operation. The success of the DSD system has meant that in practice the targets have been comfortably met[22]. Despite the seemingly successful outcome of the DSD system in terms of meeting recycling targets, the system has come in for some criticism. The main criticism is that it has been focused on collection rather than on market demand and therefore huge quantities of polymers were collected before the infrastructure had been developed to recycle and re-use the materials appropriately[24]. Amendments to the 1991 ordinance were made in 1998 and involved increasing the percentage of polymers collected which must be mechanically recycled from 24 % to 36 %. The new targets defined by the Bundesrat are approximately in line with the recovery levels currently being achieved in Germany[21].

Italy

In Italy, the approval of a new recycling law (the Legislative decree n.22 of 5 February 1997: 'Decree Ronchi') promotes the concept that 'whoever produces and distributes goods

Key Facts
● The Eco-Emballages system in France recycles packaging in a similar fashion to the German 'green dot' system.
● The German DSD organisation collects and recycles domestic waste on behalf of contributing companies.
● In Italy, whoever produces and distributes goods is responsible for their recycling, and a 10 % tax on raw materials for plastics funds recycling.

that lead to wastes is responsible for their correct recovery and/or discharge as well'[25]. Italy operates a funding scheme (Replastic) whereby a fund is provided from which a clearing house can make the required handling payment (Gate Fee) to the relevant firm in the recycling chain[25]. Since 1990, Replastic, a deposit organisation for recycling plastic bottles, has been operated by an industrial consortium of plastics producers, importers, agents and users of plastic bottles. An Italian State tax of 10 % on the total value of the plastics raw material is levied on all plastics containers that are either produced in Italy or imported. This tax is collected by Replastic and used to fund the purchase of material collected and sorted by communities for recycling. For instance, in 1993 a total of 15 000 tonnes of plastics were collected in 1500 towns at a cost of 200 lire *per* kg of bottle waste[25]. Italy also has a national packaging consortium, Conai, made up of commercial companies, which pay more than 207 million euros to the consortium. Conai, in turn, collects and recycles packaging materials. Recent collaboration between Conai and the association of Italian municipalities (ANCI) will in the future provide a means to improve segregation of waste and thus increase recycling rates.

The Netherlands

In 1991, a covenant was drawn up between the government and the packaging industry with a duty on the packaging industry that they would voluntarily set up the necessary infrastructure to ensure viable collection, sorting and recycling systems by January 2001[26]. A further covenant was agreed in December 1997 following the introduction of the Packaging and Packaging Waste Regulation earlier that year. In 1991, the covenant also stated that this objective had to be achieved by at least 10 % prevention and at least 65 % material recycling of the total quantity of packaging placed on the market in 2001.

VMK, the Dutch Association for Environmental Management of Plastics Packaging, is the environmental management organisation for the packaging industry in the Netherlands[27]. The group (currently comprising 250 members) seeks to promote the interests of companies engaged in the supply of raw materials for plastics packaging, the production and sale of plastics packaging and the collection and recycling of plastics packaging waste. One of the aims of VMK is to liaise between the plastics packaging sector and government agencies, environmental organisations and consumer organisations. All Dutch companies in the packaging or packaged goods sectors are subject to the Dutch government's Packaging and Packaging Waste Regulation. For any company, complying with this regulation individually would mean a considerable financial and administrative burden, but under the terms of Packaging Covenant II (December 1998) companies can seek collective compliance (as VMK members).

Portugal

Portugal is also an advocate of the 'green point' system[21]. Raw materials producers and developers are responsible for the utilisation of the packaging waste fraction in MSW. Commercial and industrial packaging distributors and users are responsible for utilisation of these products and this must be carried out in a regulated way, or within an integrated system. In common with Denmark, the use of refillable bottles is obligatory for bottling beer and all carbonated beverages and retailers who introduce products in non-refillable containers must offer the same product categories in refillable packaging. In the case of non re-usable packaging, the Portuguese Government places the responsibility for post-consumer management squarely on the producers, transformers and users, by insisting that the organisations concerned must create their own packaging recovery system, or participate in an integrated system.

Spain

Spanish product manufacturers, retailers and local authorities have been developing a plan of voluntary agreement for packaging wastes similar to the French 'green point' system.

Key Facts ⚹━━ᛟ
● The Netherlands, Portugal, Spain and Turkey all have 'green point' type systems which place recycling responsibility on the producer and fund recycling schemes.

Producers and distributors are given certain obligations for recovery of their products. The Government establishes annual limits of recovery (after consultation with regional authorities and industry), except for refillable containers and the producers and distributors are exempt from these obligations if they join an integrated recovery system. Additional costs (*i.e.* the difference between differentiated and undifferentiated collection) are paid by each industry with a system based on the 'green point'. Among the objectives set is the reduction by 20% of the use of PVC in food packaging; 'eco-taxes' are levied on packaging that fails to achieve these objectives.

Turkey

Although not yet an EU country, Turkey has relatively high recycling rates for some materials, for example as high as 40% for PE[21]. The same reference reports that the total consumption of plastics in Turkey was reported to be more than 1 million tonnes in 1994, with thermoplastics making up around 60% of the figure. Of these, LDPE, PVC, PP and PET made the highest contributions. Scraps of plastic waste generated during production are sold or used by the manufacturer, while post-consumer wastes are collected in several ways. Some items (*e.g.* greenhouse films or fertiliser bags) may be collected as part of a well organised system by large consumers and subsequently sold or, on a somewhat more *ad hoc* basis, by the smaller consumers (*e.g.* supermarkets, restaurants or households). Almost 60% of the plastic scrap in MSW is collected in this fashion, leaving the remainder to be transferred to waste fields by the municipal Government, from which it is auctioned to buyers (who then separate the waste manually). PET and PVC bottle producers occasionally organise collection campaigns to generate recyclate.

United Kingdom

The recovery and recycling targets set out in the EU Packaging Directive are intended to be met in the UK through implementation of the Producer Responsibility Obligations (Packaging Waste) Regulations 1997. These Regulations impose obligations on producers to recover and recycle packaging waste. The Regulations impose a duty on producers to register with the Environment Agency or with a collective compliance scheme and to take reasonable steps to recover and recycle packaging waste. They must then furnish a certificate of compliance in respect of their obligations to the Agency. The Regulations originally applied to producers who handled more than 50 tonnes of packaging and with an annual turnover above £5 million. As from 2000, this obligation has been extended to businesses with an annual turnover of more than £2 million where more than 50 tonnes of packaging are handled. Individual producers may register or they may join a scheme whereby the operator of the scheme undertakes that the obligations of all the members in relation to recycling and recovery will take place through the scheme. The Regulations are being implemented through a system of 'Packaging Recovery Notes' (PRNs) which has been devised by the Government so that the scheme will be subject to market forces. Companies who reprocess waste materials in a production process for the original purpose or for other purposes, are able to issue a PRN.

In order to achieve the EC packaging targets, the UK government concluded that businesses falling under the Regulations would have to recover 52% of packaging and 16% of the main materials. It would appear, however, that these targets are not sufficient to achieve compliance with the EC Directive and the targets are again under review. In particular, it is proposed that a specific recycling target of 18% for obligated businesses is required to achieve the national target of 15% for plastics. Mechanical reprocessing of the recovered polymer is an option as well as chemical or feedstock reprocessing.

Recovery and recycling of plastics in the UK are promoted through a number of organisations and projects. RECOUP (Recycling of Used Plastic Containers, Ltd.) is one such non-profit making company, which aims to promote and facilitate the recycling of

Key Facts
● In 1997, the UK introduced regulations that made producers responsible for waste recycling.
● The regulations are enforced using 'packaging recovery notes' (PRNs) issued by reprocessor companies.

post-consumer plastic in the UK. It is funded by companies involved in the manufacture, filling and use of plastic containers, along with trade bodies such as the British Plastics Federation (BPF). RECOUP provides financial support and guidance in order to promote each element of the recycling chain, *i.e.* material, collection, cleaning and separation, and reprocessing of materials into new uses.

RECOUP works in collaboration with LINPAC-Environmental, the largest recycler of rigid plastic in the UK. LINPAC-Environmental is involved in recycling numerous products such as vending machine cups, bottles and processes around 10 000 tonnes of recycled material each year. Many of the products it produces are made totally from recycled material. LINPAC-Environmental[28] buys recovered material obtained through RECOUP.

Wellman International is another company that works in collaboration with RECOUP. It is Europe's largest recycler of plastics overall and is particularly involved in the production of polyester fibre from recycled material[29]. Wellman is based near Dublin, Ireland and has additional recycling sites in France and the Netherlands, which together process 730 million plastic bottles. In 1995 the company worked together with RECOUP on a project funded by the EC LIFE Programme to look at ways in which plastic bottle recycling could be improved in the north west of England. By 1997 it had been found[30] that over a period of 8 years, the amount of post-consumer plastic bottle material recycled in the UK had risen significantly from 340 tonnes in 1990 to 7000 tonnes in 1997. However, projections based on current recycling rates and local authority plans suggested that additional assistance would be required for the 2000 target to be met.

A further activity related to plastics recycling is known as 'CARE' (The Consortium for Automotive Recycling). This is a project involving UK vehicle manufacturers, importers and dismantlers that carries out research into the re-use and recycling of material from scrap vehicles. The CARE group also works closely with ACORD (Automotive Consortium on Recycling and Dismantling), which was established by the UK Car Manufacturers in order to develop and implement a multi-industry strategy to improve the existing disposal processes for vehicles. The ACORD agreement contains a commitment by the trade associations involved to improve the recovery of material to 85 % by 2002, and 95 % by 2015. To do this they are involved in developing new processes to make recovery of non-metallic materials more effective[29].

In addition, there are several non-profit organisations within the UK which are involved in promoting the principles of waste reduction, re-use and recycling and which aim to bring together industry, local authorities, private and public sector institutions to develop recovery and recycling schemes. For instance, 'Save a Cup' recycling company is a non-profit company involved in collecting EPS vending cups from businesses and recycling them into products such as pencils, rulers and bins. It has developed an 'environmental charge' on all new vending cups to help fund their initiatives. To date, they have collected and recycled more than 1 billion cups[31]. Other non-profit organisations include Valuplast, Waste Watch and The National Recycling Forum.

3.5.2 North America

United States of America

With a population of at least 250 million people[32] and a highly developed consumer economy, the need to implement a robust solid waste management scheme in the USA is crucial. With the production of polymers in excess of 30 million tonnes *per annum* (1995 data), the country is the world's largest polymer producer. The amounts of plastics consumed annually in the USA have grown steadily from 1960, when they constituted 0.5 % of the MSW, to 1996, when the figure had risen to 12.3 %[32]. The five most prevalent polymers in MSW are (in decreasing order) LDPE, HDPE, PP, PS and PET. Fortunately, post-consumer plastic recycling has also grown appreciably in recent years; *e.g.* 617 000 tonnes of rigid plastic containers were recycled in 1997, representing a growth of around 4.1 % over the previous year.

Key Facts
● 7000 tonnes of plastics were recycled in the UK in 1997.
● America produced 30 million tonnes of plastic bottles in 1995; 0.6 million tonnes of rigid plastic containers were recycled in 1997 using techniques such as returnable deposits on bottles, disposal bans and mandatory coding of polymer types.

Most states in the USA have established waste diversion or reduction goals (generally voluntary in nature), but these vary significantly from state to state. Regulation may take the form of:

- Forced deposits, ranging from 5–10 cents, are imposed on certain consumer beverages (*e.g.* PET bottles). The sum is paid by the consumer, who can then redeem the deposit by returning the empty container to an established redemption centre. The price of the beverage also includes a handling fee to support the redemption programme (as do unredeemed deposits).
- Disposal bans may be imposed on selected waste materials (*e.g.* on motor vehicle tyres or batteries or nonbiodegradable grocery bags) to prevent them from going to landfill.
- Restrictions on rigid containers imposed by several states specify that these must comply with different criteria (*e.g.* that they must contain a specified level of recyclate, or must be re-usable or re-fillable, in some cases for a specified number of life cycles).
- Mandatory coding of plastic bottles is mandated by the majority of the states, whereby plastic bottles of at least 16 ounces (0.45 kg) or other rigid plastic bottles of at least 8 ounces (0.23 kg) must carry an SPI resin code (recycling sign imprinted on the product to distinguish different types of polymers; these codes can be found in Chapter 4).
- Financial aid to municipalities and local public initiatives to promote recycling and waste diversion.

A report prepared by the American Plastics Council (APC) for the IUPAC working party on recycling of polymers[33] examined the 15 000 communities in the USA that participated in residential plastic recycling programmes in 1994. The APC reported that during this period over 21.3 % of all plastic bottles and 17.2 % of all rigid plastic containers found in MSW were recycled. The collection programme was divided into the following broad categories: kerb-side collection, deposit containers and drop-off programmes.

Key Facts
- The USA uses kerbside collection, deposit containers and drop-off programmes to collect domestic plastic waste.

Kerb-Side Collection

The APC estimated that there were over 9000 kerb-side collection programmes in existence in the USA in 1994[33]. These covered a variety of approaches that could be categorised in Table 3.8.

Table 3.8 Types of residential collection programmes in the USA.

Collection programme	Method of operation
Kerb-side collection of source separated material	Recyclables are deposited in different containers at the kerb-side and loaded in separate bins on the collection vehicles.
Kerb-side sorting of commingled materials	Commingled recyclables are deposited and sorted at the kerb-side into separate lorry compartments.
Commingled programmes	(a) Recyclables are set out commingled at the kerb-side and separated into fibre and containers at the lorry. (b) Recyclables are left commingled at the kerb-side and sorted on the lorry between stops.
Co-collection programmes	(a) Containers and fibres are collected in separate compartments on the lorry and MSW is compacted in a third compartment on the lorry. (b) Recyclables are set out at kerb-side in plastic bags, collected, and compacted with refuse.

Deposit Containers

These are dropped off by consumers at grocery and beverage stores and at special redemption centres established to handle this material. The general procedure may involve the beverage producer or distributor collecting the spent containers when delivering the next consignment of beverages or the involvement of a third party. The distributor or bottling company bear the cost of the collection programme and this is wholly or partially offset by a handling fee (paid to the distributor by the retailer) or unredeemed deposits. In some areas container return rates are reported to be very high, *e.g.* 90 % in Massachusetts[33].

Drop-Off Programmes

Often used in rural communities to reduce costs as residents are responsible for bringing recyclables to the drop-off facility, but costs were found to be higher than anticipated if the resident did not combine a trip to the drop-off facility with another errand. As a result the facilities were generally located at destinations which residents would already visit (*e.g.* shopping centres, schools, *etc.*). For instance, the APC study[33] showed that residents in Vermont had driven between 7000–8800 miles *per* year to drop-off facilities, compared with trips of 1250–1390 miles *per* year to kerb-side collection routes operating fortnightly.

Canada

In 1995, 3.2 million tonnes of polymers were produced in Canada[32] with the bulk of production being in the Western Canadian provinces of Alberta and Ontario. For the same year, the consumption of polymers amounted to 2.5 million tonnes with LDPE/LLDPE, PP, PVC, HDPE and EPS/PS making up around 74 % of the total (and packaging made up the largest single category at 34 %).

The Canadian Provinces and Territories are responsible for the management of nonhazardous waste. The National Packaging Protocol, which was formulated in 1989 and applied to all ten Canadian Provinces, set out to divert 50 % of the packaging waste from landfill or incineration by the year 2000; this target had already been exceeded by 1996 (51.2 % having been achieved)[32]. Despite this, no province actually stipulates a recovery rate for plastic products.

The majority of the current regulations that affect plastics take the form of deposit-return systems imposed on beverage containers (as we discussed in the case of the USA). With the exception of Ontario, which regulates only refillable *glass* soft drink containers and beer containers, some form of deposit-return system (differing in the amount of deposit that is returned) operates in all of the other provinces and territories. In Ontario, all communities numbering more than 5000 residents have kerb-side or depot collections for post-consumer PET bottles and selected nonpolymeric materials.

Quebec is one of the smaller provinces and Riedl[34] estimates that the volume of post-consumer recycled plastics in Quebec has probably topped 10 000 tonnes and, based on a consumption figure of 145 000 tonnes, he calculates a recycling rate of about 7 %. One of the principal areas of application is the development of 'plastic lumber', a timber and concrete substitute based on recycled PE/PP blends. Balatinecz and Sain[35] based in Ontario, estimated that plastics make up around 7 % (by weight) of the MSW.

3.5.3 The Pacific Rim

Australia

For Australia's largely urban population of 17 million, waste collection is predominantly the responsibility of local government[36]. Currently post-consumer waste is almost entirely sent to landfill and, in common with figures quoted elsewhere in this chapter, plastics make up a relatively small component (*ca.* 7 % by weight) of Australian MSW. There are several reasons for Australia's traditional reliance on landfill. In Truss and O'Donnell's view[36] some costs associated with landfill (*e.g.* the deterioration in environmental amenity; the environment contingency costs; closure and post-closure monitoring costs; labour and equipment; and the cost of the land used for the landfill) have been undervalued in the past. Given the remote location of the sparse population in some areas, the low volume of recyclable material and cost of transport, landfill has been seen as the preferred option.

Key Facts 🔑

● In 1995, Canada consumed 2.5 million tonnes of plastics and now operates deposit return and kerbside collection schemes to recycle plastics.

● Australia has plenty of spare land and sends most of its MSW to landfill.

The Australian plastics market is reported to be around 1 million tonnes *per annum*[37], of which 70 % is consumed in long life products (*e.g.* cars, buildings, electrical goods, *etc.*) with a useful life of greater than 5 years. The remaining 30 % are consumed by the packaging industry and appear in the waste stream. In common with other geographical areas already covered, the bulk of the waste stream is made up from six major commodity polymers (Table 3.9) in an approximate order of tonnage LDPE>HDPE>PP>PET>PS>PVC. Of the plastics represented in Table 3.9, PET stands out significantly from the others in terms of the amount recovered, primarily because it is used in soft drinks bottles. PET is a particularly interesting example to examine in this context since there is no producer of this polymer based in Australia and all of the PET used in 1992 (21 000 tonnes) was imported into the country.

Table 3.9 Quantities of plastics collected in Australia in 1992[36]. Reproduced with permission from Truss and O'Donnell (1998). Plastics recycling: an Australian overview. In *Macromol. Symp.*, **135**, 345–358. Copyright (1998) Wiley-VCH, Weinheim.

Polymer	Amount consumed (tonnes/year)	Amount recovered (tonnes/year)	Recovered (%)
PET	28 000	5435	20
HDPE	140 000	4612	3.3
PS	46 000	652	1.4
PVC	170 000	400	0.2
LDPE	121 000	713	0.6
PP	160 000	385	0.2

A Strategic Industry Research and Analysis report[37] written in 1994 indicated that 47 000 tonnes *per annum* of plastic waste are recycled in Australia from industrial and commercial sources, along with 12 000 tonnes *per annum* from post-consumer domestic waste. However, while the rate of recycling operated by industrial sources was reported to have remained quite stable through the 1990s, a marked growth in the rate of post-consumer waste collection had been observed during 1989–1994. These data confirmed the increase in public perception and awareness of the need to recycle in Australia. For instance, the successful Sydney bid for the year 2000 Olympic Games was billed as the 'Environmentally Friendly Games' and this theme directly influenced the choice of building materials, biodegradable packaging materials and crockery and cutlery associated with the event[36].

Ambitious targets have been set: by 2000, the Commonwealth Environmental Protection Agency (CEPA) was aiming to reduce the total MSW by 50 %, while the Australian and New Zealand Environment and Conservation Council (ANZEC) were aiming to reduce packaging by the same amount within the same timescale. KEMCOR, the supplier of 90 % of the HDPE used for milk packaging in Australia, is trying to reduce the amount of HDPE sent to landfill by 50 %[37].

Japan

In 1993 Miki and Oki[38] reported that the plastic packaging market in Japan was experiencing steady and strong growth at the expense of metal cans and glass bottles. The reported growth was such that the proportion of plastic packages in the total packaging market was increasing by approximately 1 % *per annum*. In 1988, over 2.7 million tonnes of plastic packages were shipped within Japan, but by 1990 this had risen to 3.2 million tonnes so that plastics made up 14.8 % of all packaging materials. The major polymers employed in this way were LDPE (27.5 %), PS (22.5 %), PP (19.0 %), HDPE (17.6 %), rigid PVC (4.7 %), PET (4.5 %), flexible PVC (4.1 %) and polycarbonate (PC) (0.1 %). In Japan, polymer recycling is still very limited, the exception being PVC sheets used for agricultural purposes (of which 37 400 tonnes, or 40 %, were recycled into civil engineering materials) and foamed PS (of which 7874 tonnes were collected from high volume uses, such as fish boxes)[38].

Key Facts
● The majority of MSW is incinerated in Japan because space for landfill is scarce and expensive.

Compared to the other countries discussed so far, Japan has taken another route to waste management. The energy crisis of 1975 stimulated much research into the effective use of energy, particularly in the packaging industry. Japanese legislation ensures that MSW produced by citizens is to be collected and properly disposed of by the municipal government financed by citizens' taxes, while industrial waste must be treated by the industry itself[38]. Owing to the extremely high density of population within the Japanese islands (over 126 million Japanese inhabit an area of 377837 km^2, an average of 335 people *per* km^2), landfilling is not a viable option for disposal. Consequently, the majority of MSW (75%) is incinerated[39]. This can be in sharp contrast with countries in Western Europe where incineration either with or without energy recovery may be less widely practised. For instance, this may range from Greece, Ireland and Portugal, where no MSW is incinerated, to Italy, Spain and the UK, where around 10–18% is disposed of in this manner, through to Luxembourg and Switzerland where more than 70% of waste shares this fate[40].

3.5.4 Asia

India

Although the consumption of plastics *per capita* in India is comparatively low, it rose ten-fold during the period 1986–1995 to reach 1.8 million tonnes[41]. This represented *ca.* 1.8 kg *per* person yearly (the world average is *ca.* 18.0 kg *per capita-annum* and the developed world average is 80 kg *per capita-annum*), but the figure was projected to double[41] by 2001 to reach 3.5 million tonnes (2.9 kg *per capita-annum*). The same researchers predicted that at the current rate the consumption of plastics in India would outstrip that of paper by around 2005. In 1998 the same researchers reported that the collection, transportation and disposal of MSW in India were 'unscientific and chaotic'. Citing a 1996 report by the Tata Energy Research Institute, the authors stated that plastic waste constituted 4–9% of MSW, depending on income level. Some forms of plastic waste (*e.g.* bottles and jars) are segregated by the householders and then sold to itinerant waste buyers, whilst packaging waste (representing around 52% of the plastics consumption), coloured polyethylene bags, *etc.* are generally discarded with the rest of the MSW. The 'waste pickers', 'ragpickers' or 'scavengers' salvage these items and through a chain of several other middlemen, the salvaged waste reaches the plastics recycling units. In 1998, India had around 2000 plastic recycling units with a production output of around 323210 tonnes (around 37.2% of the total thermoplastics in the country, including imports). Unfortunately, the authors also drew attention to the outdated technology associated with the polymer recycling units, low funding and poor quality of the raw material[41].

3.5.5 Africa

Tanzania

Information on recycling activity in the continent of Africa is scarce and it is not our intention to generalise on the basis of a single study, but the case of Tanzania does serve to highlight some interesting points. Kaseva and Gupta[42] studied the city of Dar es Salaam in Tanzania between 1993 and 1995 to examine scavenging activities and recycling trends. They reported that plastics made up around 1.9% by wet weight of MSW and around 0.48% of the total generated waste being recycled. In general, the method of materials sorting and collection bore a striking resemblance to the Indian study that we discussed earlier. Materials that were sorted out by scavengers were sold either to petty vendors and/or small-scale manufacturers located in different parts of the city, or to big industrial units through local middle-men (although the scavengers ran the risk of being exploited in this case). The study also revealed the re-use of the plastics: plastic bottles were generally used for storage purposes, while plastic sheets were used as rough covering materials.

Collection of polymer waste in different countries and factors that drive or constrain it are further elaborated in Chapter 5.

Key Facts
- In India, plastics consumption *per capita* was only 1.8 kg *per* person *per annum* in 1995.
- Plastics are recycled by 'waste pickers' and 'scavengers' who sell waste to recycling plants.

3.6 SUMMARY AND LEARNING OUTCOMES

We have tackled a number of different issues within this chapter ranging from production and application of individual polymer types through consumption to waste management at the ends of their useful lives. By reading the contents of this chapter and attempting the revision exercises, you should be able to discuss:

- how selected common industrial commodity polymers are prepared and polymerised and the physical and mechanical properties that may be developed;
- the general size of the world polymer market and the consumption of commercial polymers by each market sector in different geographical areas;
- the different sources of polymers in the environment, and their primary and recycled uses;
- the identity of the different waste stream categories and the nature of the polymers that make up each category;
- why polymers become mixed from both a thermodynamic and entropic point of view; based on your reading of Chapter 2, you should be able to suggest how the compatibility of polymer blends might be improved;
- how different countries address the problem of the collection of plastic waste.

3.7 REFERENCES AND FURTHER READING

1. Billmeyer, F.W., Jr. (1984). *Textbook of Polymer Science*, 3rd edn, John Wiley, New York, pp. 364–366.
2. Hannig, N. (1995). Properties and applications of recycled PE-HD. In *Recycling and Recovery of Plastics*, eds. Brandrup, J., Bittner, M., Menges, G. and Michaeli, W., Hanser, Munich, pp. 571–572.
3. Robinson, S. (2001). Breaking the mould, *Chem. Ind.*, 18 June, 377–378.
4. Seiler, E. (1995). Properties and applications of recycled polypropylene. In *Recycling and Recovery of Plastics*, eds. Brandrup, J., Bittner, M., Menges, G. and Michaeli, W., Hanser, Munich, pp. 599–614.
5. Walter, H.-M. (1995). Properties and applications of recycled polystyrene. In *Recycling and Recovery of Plastics*, eds. Brandrup, J., Bittner, M., Menges, G. and Michaeli, W., Hanser, Munich, pp. 615–630.
6. APME (2001). *An Analysis of Plastics Consumption and Recovery in Western Europe 1999*, APME, Brussels. Also available at: http://www.apme.org/dashboard; 20 June 2002.
7. APME (1997). *Plastics consumption and recovery in Western Europe in 1995*, APME, Brussels.
8. Blumschein, H.-W. (1995). Properties and applications of recycled poly(ethylene terephthalate). In *Recycling and Recovery of Plastics*, eds. Brandrup, J., Bittner, M., Menges, G. and Michaeli, W., Hanser, Munich, pp. 669–675.
9. Kresta, J.E., Xiao, H.X., Suthar, B., Li, X.H., Sun, S.P. and Klempner, D. (1998). New approach to recycling of thermosets, *Macromol. Symp.*, **135**, 25–33.
10. Boustead, I. (1996). *Polyurethane Precursors (TDI, MDI, Polyols), Report 9. Eco-profiles of the European Plastics Industry*, APME and ISOPA, Brussels, June.
11. Bükens, A.G. and Schöters, J.G. (1998). Technical methods in plastics pyrolysis, *Macromol. Symp.*, **135**, 63–81.
12. Stein, R.S. (1998). Polymer recycling: thermodynamics and economics, *Macromol. Symp.*, **135**, 295–314.
13. Matthews, V. (1995). Plastics consumption and plastics waste in Western Europe – a statistical survey. In *Recycling and Recovery of Plastics*, eds. Brandrup, J., Bittner, M., Menges, G. and Michaeli, W., Hanser, Munich, pp. 531–541.
14. Roy, R. (1991). *End-Of-Life Electronic Equipment Waste*, CEST.

15. CEST (1991). Disposal of Vehicles: Issues and Actions.

16. EC (1994). Council Directive 94/62/EC of 20 December 1994 on Packaging and Packaging Waste, *Off. J. Eur. Commun.*, No. L 365, 10–23.

17. EC (2000). Proposal for a Directive of the European Parliament and of the Council on Waste Electrical and Electronic Equipment. COM (2000) 347 final, Brussels, 13 June 2000, httl2://www.europa.eu.int/eur-lex/en/index.html.

18. Hardman, S. and Wilson, D.C. (1998). Polymer cracking – new hydrocarbons from old plastics, *Macromol. Symp.*, **135**, 113–120.

19. APME (1998). *Options for Plastics Waste Recovery*, December, http://www.apme.org/environment/htm/06.htm, accessed on 5 July 2001.

20. PIRA (2001). Evaluation of costs and benefits for the achievement of re-use and recycling targets for the different packaging materials in the frame of the Packaging and Packaging Waste Directive 94/62/EC, Proposed Draft Final Report. RDC, Environment & Pira International, May, p. 341.

21. Curto, D. and Basar, Y. (1998). Regulations and practices of polymer recycling in NATO countries A – European countries, Chapter 1: Introduction. In *Frontiers in the Science and Technology of Polymer Recycling*, eds. Akovali, G., Bernardo, C.A., Leidner, J., Utracki, L.A. and Xanthos, M., Kluwer, Dordrecht, Netherlands, pp. 17–28.

22. Smosarski, G. (1995). Materials recycling – turning waste into valuable materials. In *Financial Times Management Report*, Pearson Professional.

23. Brandrup, J. (1998). Ecological and economical aspects of polymer recycling, *Macromol. Symp.*, **135**, 223–235.

24. Johnson, D. (1994). Plastics the environmental challenge. In *Financial Times Management Report*, Pearson Professional.

25. Bruder, J. (1995). Economic instruments for encouraging plastics recycling. In *Recycling and Recovery of Plastics*, eds. Brandrup, J., Bittner, M., Menges, G. and Michaeli, W., Hanser, Munich, pp. 46–60.

26. Koca, D. and Nilsson-Djerf, J. (2000). Assessment of different methodologies/studies used to determine the amount of packaging waste that remains in final disposal – experiences from six European countries, Lund University Centre for Applied System Dynamics (LUCAS).

27. VMK – Dutch Association for Environmental Management of Plastics Packaging, http://www.vmk.nl, accessed 20 June 2001.

28. LINPAC Environmental, http://www.linpac-environmental.com/, accessed 19 June 2001.

29. Wellman International Environmental, http://www.wellman-intl.com, accessed 19 June 2001.

30. Smith, D.N., Harrison, L.M. and Simmons, A.J. (1999). A survey of schemes in the United Kingdom collecting plastic bottles for recycling, *Resour. Conserv. Recycling*, **25**, 17–34.

31. CARE, http://www.caregroup.org.uk/index.shtml, accessed 19 June 2001.

32. Edgecombe, F.H.C. (1998). Regulations and practices of recycling in NATO countries B – Canada and United States of America, Chapter 1: Introduction. In *Frontiers in the Science and Technology of Polymer Recycling*, eds. Akovali, G., Bernardo, C.A., Leidner, J., Utracki, L.A. and Xanthos, M., Kluwer, Dordrecht, The Netherlands, pp. 29–39.

33. Subramanian, P.M. (2000). Plastics recycling and waste management in the US, *Resour. Conserv. Recycling*, **28**, 253–263.

34. Riedl, B. (1998). Recycled injection moulded and fibre reinforced poly(ethylene-propylene), *Macromol. Symp.*, **135**, 121–128.

35. Balatinecz, J.J. and Sain, M.M. (1998). The influence of recycling on the properties of wood fibre-plastic composites, *Macromol. Symp.*, **135**, 167–173.

36. Truss, D.W. and O'Donnell, J.H. (1998). Plastics recycling: an Australian overview, *Macromol. Symp.*, **135**, 345–358.

37. Strategic Industry Research and Analysis (SIRA) (1994). Australian plastics materials recycling survey. Report for the Plastics and Chemicals Industries Association.

38. Miki, H. and Oki, Y. (1993). Worldwide environmental contrasts – Japan. In *Packaging in the Environment*, ed. Levy, G.M., Blackie Academic and Professional, Glasgow, pp. 245–265.

39. Akehata, T. (1998). Energy recovery, *Macromol. Symp.*, **135**, 359–373.

40. APME (1997). *Recovery of Plastic Waste in Western Europe*, Association of Plastic Manufacturers Europe, Brussels.

41. Gupta, S., Mohan, K., Prasad, R., Gupta, S. and Kansal, A. (1998). Solid waste management in India: options and opportunities, *Resour. Conserv. Recycling*, **24**, 137–154.

42. Kaseva, M.E. and Gupta, S.K. (1996). Recycling – an environmentally friendly and income generating activity towards sustainable solid waste management. Case study – Dar es Salaam City, Tanzania, *Resour. Conserv. Recycling*, **17**, 299–309.

Further Reading

Anastas, P.T. and Warner, J.C. (1998). *Green Chemistry Theory and Practice*, Oxford University Press, Oxford.

Bakker, M. and Gighotti, M. (1993). Worldwide environmental contrasts – North America. In *Packaging in the Environment*, ed. Levy, G.M., Blackie Academic and Professional, Glasgow, pp. 212–244.

Brandrup, J., Bittner, M., Menges, G. and Michaeli, W. (eds) (1995). *Recycling and Recovery of Plastics*, Hanser, Munich.

Cowie, J.M.G. (1991). *Polymers: Chemistry and Physics of Modern Materials*, 2nd edn, Blackie, Glasgow.

Johnson, E. (1993). Worldwide environmental contrasts – Europe. In *Packaging in the Environment*, ed. Levy, G.M., Blackie Academic and Professional, Glasgow, pp. 187–211.

Kahovec, J. (ed.) (1998). Recycling of polymers, *Macromol. Symp.*, **135**.

Levy, G.M. (ed.) (1994). *Packaging in the Environment*, Blackie Academic and Professional, Glasgow.

Kirkwood, R.C. and Longley, A.J. (eds) (1995). *Clean Technology in the Environment*, Blackie Academic and Professional, Glasgow.

Lundquist, L., Leterrier, Y., Sunderland, P. and Manson, J.-A.E. (2000). *Life Cycle Engineering of Plastics*, Elsevier, Amsterdam.

Saunders, K.J. (1988). *Organic Polymer Chemistry*, 2nd edn, Chapman & Hall, London.

3.8 REVISION EXERCISES

1. Discuss the primary and secondary uses of PET, LLDPE and PVC. Your answer should address the factors that determine whether the recycling process is viable, including the ease of processing and/or sorting and the deterioration in properties during recycling.

2. PE is available in many commercial forms, depending on the degree of branching in the polymer chain. Sketch the mechanism to account for the formation of *n*-butyl branches in LDPE, showing the electron shifts. How does the degree of branching (and the nature of the branches) affect the physical and mechanical properties of PE and the uses to which the different forms are put?

3. Discuss the relative physical properties of PS, impact-PS and ABS polymers, and account for any differences in terms of molecular structure.

4. Discuss the four principal synthetic methods used to produce HDPE. How do the polymers thus produced compare in terms of their structures, physical forms and mechanical properties?

5. Between 1980 and 1996 there was a substantial growth in the sales of PP and the world output of all forms of the polymer was *ca.* 16 million tonnes in 1996. The price

of PP fell dramatically in 1993, although the world demand for the polymer continued to grow (30 million tonnes in 2000). How do you think that this change affected the market (and the technological applications) for virgin PP and PP recylate?

6. Despite the market for thermosetting polymers being significantly smaller than that for thermoplastics, there is still a potentially useful stock of recyclate if the problems of recycling can be overcome. Which options are currently available for the recycling and re-use of thermosetting polymers and what problems do you envisage, associated with the network structure, during the characterisation and reprocessing of post-consumer thermosetting polymers?

7. Compare the different waste stream categories presented here in terms of their size and homogeneity. Do similar polymers appear in all the categories? If this is not the case, does this suggest that we use too many types of polymer? How would you alleviate this problem?

8. Considering the total post-consumer plastics waste generated in Western Europe in 1995, which were the worst offenders (per head) among the European countries? Can you see any trends between the waste stream categories for each of the European countries and the lifestyles of its inhabitants?

9. Discuss critically the different methods of residential waste collection programmes employed in the USA. How does this approach differ from the green dot system introduced in many EU countries? How is the UK approach different from the EU and USA approaches?

10. Use publicly available information to study recycling practices in your country. How do they differ from those of other countries in your geographical region and more widely? What do you conclude about data availability and reliability with respect to the recovery and recycling figures?

Chapter 4 – The Wheel of Fortune (E Burne-Jones, 1875–83)
Burne-Jones painted this image as part of an (unfin-ished) triptych on the Fall of Troy. His comment on the allegorical subject, in which Fortune turns the wheel to which are bound a slave, a king and a poet, was that: "(my) Fortune's Wheel is a true image, and we take our turn at it, and are broken upon it". The image of the wheel implies perpetual motion and links well with the theme of recycling technologies being implemented to return the "spent" polymer to a new working life.

managing polymer waste: technologies for separation and recycling

4.1 Introduction

4.2 Identification and Separation of Polymers

4.3 Technologies for Polymer Recycling

4.4 Summary and Learning Outcomes

4.5 References and Further Reading

4.6 Revision Exercises

4.1 INTRODUCTION

We have seen in Chapter 1 that any waste materials, including polymers, are best managed within an integrated waste management scheme. The most sustainable waste management option is always reduction at source, *i.e.* reduced consumption of materials and products, followed by direct re-use. At the other end of the spectrum is disposal by landfill, which wastes valuable resources, both material and energy, and increases the amount of solid waste in the environment. In the waste management hierarchy, recycling sits somewhere in between, depending on the resources used and wastes generated in the overall recycling operations, which must not exceed the environmental benefits of recycling. The latter are explored at length in Chapters 6 and 7, but in this chapter we concentrate on recycling methods and technologies available for plastic materials.

Three generic options are available for the recycling of polymers:

- mechanical recycling,
- chemical recycling,
- energy recovery.

Key Facts
- The environmental costs of recycling must not exceed the benefits.
- Materials recycling (mechanical or chemical) is the preferred recycling option in most national and international directives.

Mechanical recycling uses physical means, such as grinding, heating and extruding to process waste plastics into new products. Chemical recycling on the other hand uses chemical processes to convert waste into useful products, such as monomers for new plastics, fuels or basic chemicals for general chemical production. These two options are often referred to as material recycling. The third recycling option, energy recovery generates heat or electricity (or both) either by direct incineration of polymer waste, *e.g.* in municipal sold waste (MSW) incinerators, or by replacing other fuels, *e.g.* in blast furnaces, cement kilns or power stations.

The recycling activity can be divided into two main steps:

- recovery of post-consumer polymer waste (*i.e.* its collection and delivery to the recycling facility);
- waste sorting and separation.

Recovery of post consumer waste is discussed in Chapters 3 and 5; here, we examine examples of sorting, separation and processing using currently available technologies for each of the three recycling options. A number of interesting technologies for the future, some of which are still at an experimental stage, are described in Chapter 8.

4.2 IDENTIFICATION AND SEPARATION OF POLYMERS

Despite intense efforts over recent years to increase the recycling and re-use of post-consumer waste plastics, the proportion that can actually be recycled is still extremely small, not least because of the many technological and technical factors that constrain recycling. One of these is the fact that some of the recycling options require prior sorting or separation of polymers by type. The reasons for this may be technical, or based on environmental or health issues. For example, mechanical recycling is only economically and technically feasible for single materials while incinerators must separate out PVC from the MSW to avoid the formation of toxic dioxins. Both these issues will be discussed further later in this chapter.

We now explore how some of these constraints can be overcome using existing and developing technologies to identify, separate and process waste polymers.

4.2.1 Identification of Polymers

Some polymers, such as PET in soft drink bottles or PE in milk and water jugs, may be relatively easy to identify. However, many polymers in use often resemble one another and are clearly difficult to differentiate, even for a polymer scientist or engineer who may be familiar with the polymer. In 1988, the Society of the Plastics Industry (SPI) proposed the Voluntary Plastic Container Coding System, a series of seven numbered classifications (Figure 4.1) to help consumers classify different plastics and to make it easier to identify the polymers in the waste stream. In practice, the code is impressed or printed on the plastic component.

Typical examples of these polymers in post-consumer waste are:

- PET: beverage containers, boil-in food pouches, processed meat packages.
- HDPE: milk bottles, detergent bottles, oil bottles, toys, plastic bags.
- PVC: food wrap, vegetable oil bottles, blister packaging.
- LDPE: shrink-wrap, plastic bags, garment bags.
- PP: margarine and yoghurt containers, caps for containers, wrapping to replace cellophane.
- PS: egg cartons, fast food trays.
- Other multi-resin containers or microelectronic components.

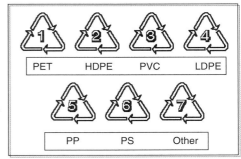

Figure 4.1 Codes used to identify important commercial polymers used in packaging

In 1991, the American Society for Testing and Materials (ASTM) issued a more general system, based on abbreviations recommended by the International Organisation for Standardisation (ISO), and encompassing over 100 polymers and polymer blends. The issue of waste stream identification is important in the automotive industry, which has been quite pro-active in exploring the recovery and recycling options. The Society of Automotive Engineers (SAE) employs a third system[1], which is very similar to the ISO model, to identify the polymers used in automotive components.

This diversity of identification systems is one of the obstacles for more effective separation of polymeric materials. The universal adoption of an industry agreed product marking system such as ISO 11469[2] for the identification of a variety of plastics would ease sorting problems during recycling. In some cases, recognition is already possible and some of the methods that are available are discussed below, but improved analytical and recognition methods are needed.

Separation is further complicated by the current design of products in which polymers are commingled with other materials. Disassembly and separation of materials can be simplified by appropriate product design and choice of materials. Various 'Design For the Environment' (DFE) approaches have been developed to assist designers and engineers in designing and manufacturing more sustainable products. DFE is discussed in more detail in Chapters 6 and 7.

Key Facts
- Polymers are difficult to identify unless labelled at source, but there are currently three labelling systems in use.

4.2.2 Sorting and Separation of Polymers

Polymers are currently sorted and separated either manually or mechanically. Considerable work has been undertaken to develop automated technologies that can separate mixed waste streams according to polymer type or to remove foreign matter (contaminants) from the waste stream. While some of these techniques are becoming quite sophisticated, in some cases the process is still in its infancy and more development work is required before the technique may be used on anything greater than a laboratory scale.

The choice of the appropriate technique will depend on a number of factors, including:

- complexity of the polymer mixture, *i.e.* the number of polymers present and recoverable;
- physical form of the polymer(s), *i.e.* whether the material is in bulk or granulated;
- quality, *i.e.* acceptable level of contamination in the recyclate for future use;
- nonplastic contaminants, *i.e.* the nature and concentration of additives;
- market, *i.e.* the eventual use to which the recyclate will be put;
- economics, *i.e.* the scale and cost of the separating operation and whether the scale is sufficiently large to justify the more expensive automatic detection and sorting methods.

Key Facts ✏️ ✎
● Pre-separation by consumers is not always reliable.
● Chemical identification of polymers often needs several techniques in combination, which is expensive.

To date, most of the development work has been carried out to separate mixed plastic bottles and containers into clean fractions of their main polymeric components and these will be discussed in greater detail. You can see from Table 4.1 that many techniques are being used or examined for the detection and separation of whole bottles and containers from the waste stream. For instance, bottles may be collected by consumers (either having undergone some sorting by polymer type or simply collected as mixed plastic waste) or by contractors (as mixed recyclable material or within MSW). If some pre-sorting has already taken place (perhaps through a 'kerb-side' or 'bring' scheme), then the materials may be fed directly into a recycling operation (although some further sorting will be necessary as sorting at source, *i.e.* by consumers, is not always reliable). These materials can be shredded to potentially yield a clean flake with minimal treatment (*e.g.* to remove contaminants such as labels or lids, *etc.*) prior to granulation and cleaning. If this is the case, then a selection of techniques is already available for the detection and separation of granulated or polymer flake (Table 4.2). In some of these cases, recognition of different polymer types is already possible on a commercial scale but, as already mentioned, there is a need for improved analytical and recognition methods.

The following methods can be used for polymer recognition:

● infrared spectroscopy or mass spectroscopy (specific chemical groups);
● thermal analysis, *e.g.* differential scanning calorimetry (energy changes such as, decomposition, melting or thermal transitions);
● nuclear magnetic resonance (molecular structure);
● opto-mechanical pattern recognition (based on size and form);
● X-ray fluorescence (presence of metals or hetero-atoms).

The efficiencies of Fourier transform infrared (FTIR) and Fourier transform Raman (FTR) spectroscopies have also been tested. Although it was possible with both techniques to

Table 4.1 Detection and separation techniques for whole bottles and containers.

Method	Mode of operation	Advantages	Disadvantages
Manual separation	Feedstock passed along a conveyor belt for operator to identify visually, sort by polymer category and separate articles.	Cheap in terms of capital investment (no sorting technology employed).	Labour intensive and slow (*ca.* 50–200 kg h^{-1} depending on number of operations). Potentially inaccurate (*ca.* 80–95 % accuracy) at economic speeds (6–10 bottles *per* second).
Manual separation assisted with some degree of automation	Feedstock passed along a conveyor belt for operator to identify visually and sort by polymer category by activating automatic ejection mechanism.	Relatively simple to blow/push bottles/containers into side chutes using basic electronics.	Less labour intensive but still slow and prone to errors.
Automatic bottle sorting according to polymer type	X-ray detection is used to detect chlorine in PVC with mechanical removal from waste stream. PE is separated from PET using optical sensor (differentiates opaque, translucent, transparent bottles) or near infrared spectroscopy.	Detection accuracy can be very high (lower than 1 error in 10 000). High throughput achievable (1 tonne h^{-1}).	High capital costs. Requires good quality feedstock and accurate presentation of PVC bottles if high accuracy/throughput to be achieved. Errors usually due to overlapping bottles or smaller units (<5 cm).
Polymer markers	Addition of markers or tracers added during manufacture: infrared dyes in PET or fluorescent dyes.	High detection accuracy for different polymers or even different grades of polymer (up to 90 % for coloured particulate contamination from clear PET flake[3]).	Much development work required. Universal acceptance of markers is essential for success. Some risk of cross-contamination and mixing.

recognize most usual polymers, severe limitations of FTIR were experienced, including a high sensitivity to the surface state. FT Raman proved to be a rapid and highly selective method, giving information even on the mineral fillers present in plastic. In most cases, a combination of different analytical methods is necessary and this has serious implications for the cost of the operation.

Increasing the complexity of the polymer mixture leads to a significant increase of the costs of separation. Engstrom[3] estimates that if pre-sorting has taken place then an operation involving collection skips, baling equipment and transportation will cost of the order of DM 50 000 to 200 000 (approximately £17 000–£65 000) for a throughput of $1-2$ tonnes h^{-1}. Manual separation will raise the labour costs, but involves little in the way of capital investment. However, if substantial automatic separation becomes necessary, then the costs for the equipment can be as high as DM 200 000 to 2 000 000 (depending on

Table 4.2 Detection and separation techniques for granulated or flaked polymer material (selected data collated from ref. 3).

Method	Mode of operation	Advantages	Disadvantages
Flotation tanks (Reprise Technologies)	Separation in liquid by means of differences in polymer densities.	Can effect coarse separation. Improvements are being examined with supercritical fluids and other non-organic solvents to replace water.	Needs at least two stages for good separation. Problems encountered if polymers have very similar densities.
Hydrocyclone (Byker reclamation plant, Newcastle, UK)	Pressurised separating fluid (usually water) containing polymer particles forms vortex (250 times the force of gravity). Separation effected by centrifugal force.	Allows better separation than static flotation tanks. May separate PE from heavier PVC, PET or PS. Higher throughout than static float and sink tanks.	Problems encountered if polymers have very similar densities or if grain size and shape differ greatly.
Compressed air separation	Rotating drum with an air flow to fluidise light component (plastic film/paper labels/cloth) and allow separation by density.	Allows better separation than flotation tanks. May separate PE from heavier PVC, PET or PS.	Problems encountered if polymers have very similar densities.
Micronising (Reprise Technologies, UK and Ecoplas, Belgium)	Pulverising and sieving polymer fractions to 600 μm, utilising the possibility that different polymers have different grinding characteristics.	PET contamination in PVC can be reduced from 2 % to 0.2 %. Cryogenic grinding improves efficiency. By freezing the samples (below T_g), the integrity of the solid samples is maintained and separation facilitated.	Economic and environmental costs of cryogenic grinding are high due to high energy requirements.
Electrostatic separation Devtech Labs. (USA) Kali und Salz Co. (Germany)	Charged polymer flakes fall between charged plates and separated based on their capacity to retain charge, which is related to bulk and surface resistivity.	Good removal of residual contamination from PVC or PET.	Critically dependent on pretreatment of flake surface to achieve a reproducible charge.
Melt filtration Herbold Co. (Germany) Gneuss (Germany)	Polymer components heated on band conveyor and a rotating drum picks up softened material. Contaminants can be removed from molten polymer by allowing the molten polymer to flow through screen apertures.	Problems may be encountered with ensuring that the filter area is continuously renewed without interrupting the process.	May be difficult to move beyond coarse scale separation.
Separation of coloured particles (Reprise Technologies, Radex Systems, Ltd., Sortex, Envirotechnics, UK)	Low technology optical detection and air jet removal; or laser detection and air jet removal.	Optical discrimination (Reprise) effective to remove greater than 90 % of coloured contaminants.	Very low levels required difficult to achieve for PET using optical methods.

the sophistication of the technique and the number of sorting operations required). He also cites the cost of a modern, materials' recycling facility (MRF) (to handle the separation of plastic bottles from other recyclable materials, such as paper and glass) as DM 2–20 million for a medium sized town. Finally, automating the separation of polymers from MSW would require additional effort to implement feedstock recovery or incineration and Engstrom[3] estimates the cost of such a facility as at least DM 100 million (~£35 million). Sorting and recycling costs are further discussed in Chapter 5.

4.3 TECHNOLOGIES FOR POLYMER RECYCLING

4.3.1 Mechanical Recycling

Separation of different polymers is particularly important for mechanical recycling because processing mixed materials would otherwise produce recyclate of low quality, which could only be used in a limited number of applications. Hence, mechanical recycling is really best suited to clean plastic waste, such as packaging material.

Mechanical recycling of mixed waste necessarily starts with a manual sorting process, which means high labour costs and the separation is not always as efficient as it needs to be to produce high quality products. Thereafter, depending on the polymer type, the processing involves:

- Thermosets: some form of grinding and particulation for re-use as filler for new materials to improve properties such as the modulus, elongation-at-break or impact strength.
- Thermoplastics: re-melting and extrusion into new products or pelletising to be sold on as a raw material for further processing.

The following sections describe some of the processes developed for mixes wastes and for specific polymer types.

Mechanical Recycling of Mixed Plastics

A number of manufacturers are developing designs for a fully automated mechanical recycling plant, which eliminates the need for manual sorting. One such process has been developed and tested in pilot plant trials at the RWTH University of Aachen in Germany[4] and the layout of this process is shown schematically in Figure 4.2. The system accepts bags of waste and empties them on to a conveyor. A magnetic separator suspended above the conveyor removes magnetic materials first, then fibrous material is pulped and cleaned in a specially designed unit. The time material spends in the pulping unit varies between

Key Facts ✂
- Mechanical recycling is best suited to clean waste streams such as packaging, mixed wastes must be pre-sorted.
- Thermosets are ground up for use as fillers; thermoplastics are pelletised or extruded into new products.

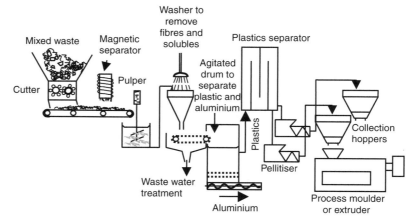

Figure 4.2 Process diagram for a plastic preparation and compounding unit DLR (German dual system) and BNIBF (German Ministry of Education, Science, Research and Technology) Process[4]. Reproduced with permission from J. Kabovoc (ed.). Recycling of Polymers, *Macromolecular Symposia*, **135**, 1998. Copyright (1998) Wiley-VCH, Weinheim

5 and 20 min, depending on the quantity of composite material in the feed and the time taken to dissolve the parts that are less soluble in water.

After the pulping process is completed, the whole contents are fed into the washing drum. Here, the dissolved or suspended paper fibre content flows through a screen to the process water cleaning unit. The remaining material in the drum contains dirt, metal and heavy, noncellulosic fibres and is washed with process water before being discharged from the drum. An agitator vessel then separates aluminium from plastics by their relative density. The aluminium, being the heaviest, settles in the sediments and is discharged by a screw conveyor and de-watered, after which it can be recycled.

The mixture of plastics is then transferred into a series of separators to be segregated into its different components. The chlorine-bearing PVC is separated using an electrostatic plate-type separator. The separation of PS is achieved in another density separator using a higher density-separating medium.

Each of the separated plastic materials is then ground, washed and dried. A new product is generated directly by extrusion (see Chapter 3 for an explanation of extrusion). Extruded products may be upgraded by the addition of fibres, mineral fillers, stabilisers, pigments, flowing aids, *etc.*

The process water produced in the pulping and cleaning is purified and re-circulated by a water-cleaning unit. Contaminated water is pumped into a thickener where a flocculant is added to improve settling out of solids, which are then discharged at the bottom and de-watered to be disposed of as a fairly dry cake.

The process recovers materials such as solid aluminium, aluminium–plastic composites and polyolefins very effectively, with a high yield and purity of material that can be used as secondary feed in industrial production elsewhere.

Clearly, the whole process is complex, which means that the initial capital cost of the equipment is high, but, being fully automated, the continuing running costs are then low relative to a process that requires manual intervention. It should be borne in mind, however, that automated separation requires a large scale operation to justify the investment costs and that may pose problems in providing an adequate volume of waste.

A second example which accepts pre-sorted plastic waste, free of iron, glass, stone and organic matter is the REVIVE system, invented by Renato Fornasero Development SAS Cadauta and promoted by the Commission of the European Communities[5]. Other impurities can be tolerated at a level of 5 % or less. Polymer waste is crushed, mixed, dried and injected into a vessel to be homogenised in the molten state. The molten mixture is then either compacted and shaped into a final product, extruded as consumer products or granulated and collected as pellets for further processing (see Figure 4.3).

CHAPTER 4

Key Facts
● Running costs of an automated plant are low compared to a manually operated plant, but the initial capital costs are high.
● Running an automated plant economically requires a high throughput of material and an adequate supply of waste.

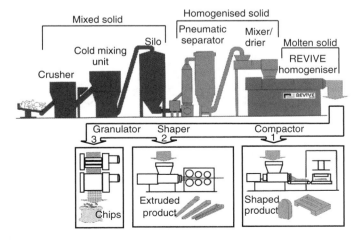

Figure 4.3 The REVIVE process for re-use of mixed plastic wastes

Recycling of Specific Polymer Types

A few specialised processes are being developed commercially for mechanical recycling of specific polymer waste sources. Two processes are illustrated here in more detail:

- selective froth floatation for recycling of PVC and PET;
- recycling of PET bottles from a homogeneous PET waste stream.

Selective Froth Flotation for PVC/PET Mixtures

As we have seen in Chapter 3, PVC and PET are two of the most used polymers. They can be reprocessed and re-melted a number of times without any significant change in their properties, due to the lack of cross-linking between the polymer chains. They have almost the same density, which makes them difficult to separate using their specific gravity. However, a selective froth flotation process has been developed by Recovery Process International, Inc.[6,7], which uses differences in hydrophobicity of the two polymers to separate and recycle them.

A simplified flow diagram of the process is shown in Figure 4.4. It is a two-step process of alkaline treatment and froth flotation and requires the size of the polymer material to be reduced to less than 10 mm to improve the efficiency of separation. The mixture is first treated with about 1–3 % sodium hydroxide for between 75 and 85 min at a temperature of about 70–85 °C, during which time the PET hydrophobicity is reduced significantly, while the hydrophobicity of PVC is only slightly affected.

After the alkaline treatment, the particles are screened and rinsed with water at a pH of 6–9. An anionic surfactant is then used at a concentration of $15–30$ mg l^{-1} for 1–2 min to generate froth. Other types of surfactant can be used, provided they have strong frothing capabilities and low affinity for the PVC and PET. PVC then floats to the top preferentially and is collected from the overflow froth over the following 5–10 min, leaving the PET behind to undergo further screening, rinsing and drying.

Laboratory scale trials on small test samples indicate that the PET recovery rate is almost 100 %, with PVC about 70 %. Most importantly, the physical properties of polymers do not deteriorate significantly during processing.

PET Bottle Recycling

The Centre for Plastic Recycling Research (CPRR)[8] in the USA has developed a process to recycle PET from a pure stream of PET bottles, which involves the following steps (see Figure 4.5):

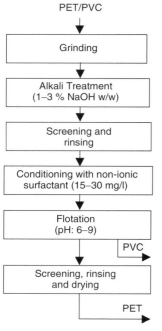

Figure 4.4 A simplified froth floatation process flowsheet for the selective recovery of PET from PVC/PET mixtures[6,7]. Reproduced with permission from '*In Polymer Engineering and Science*' v. 38/9, p. 1379–1387. Copyright (1998) SPE

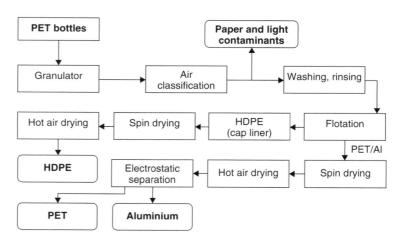

Figure 4.5 CPRR process for recycling of PET bottles[8]. Reproduced with permission from J. Kabovoc (ed.). 'Recycling of Polymers', *Macromolecular Symposia*, **135**, 1998. Copyright (1998) Wiley-VCH, Weinheim

(1) shredding the bottles to facilitate a fast granulation, followed by grinding to 1/4 inch particles;
(2) separation of paper and light contaminants in an air stream;
(3) washing and rinsing the remaining material;
(4) flotation to separate the lighter HDPE, cap liner and plastic label from the heavier PET and aluminium (from the bottle seal);
(5) separation and spin drying of lights and heavies;
(6) hot drying of each stream;
(7) electrostatic separation of PET and aluminium, where the nonconducting PET collects a charge and is attracted to oppositely charge plates;
(8) packaging and sale of aluminium and clean, uniform, granulated resins.

In summary, mechanical recycling involves the use of relatively simple processes and generates polymer materials of high quality. However, this process is mainly suitable for homogeneous waste streams which often requires clean waste of the same type or a high degree of sorting, which can increase the costs of operation. On the other hand, most of the chemical recycling processes can accept mixed or soiled waste, which are converted into feedstock for chemicals production or for use as fuel. Furthermore, the products can be separated relatively easily. The following sections outline some of the more important chemical recycling processes.

4.3.2 Chemical Recycling

Several processes can be included under the heading of chemical recycling, including gasification, hydrogenation, pyrolysis (sometimes referred to as thermolysis), and the use of waste polymer as a reducing agent in a blast furnace. The main characteristics of these processes are summarised in Table 4.3[9].

Table 4.3 Chemical recycling processes[9]. Reproduced with permission from Sasse and Emig, 'Review: Chemical Recycling of Polymer Materials.' In *Chemical Engineering Technology*, Vol. **2:1** (1998). Part 10, pp. 777–789. Copyright (1998) Wiley-VCH, Weinheim.

Process	Reactor	Reaction conditions	Products	Note
Gasification	Fluidised-bed, fixed-bed, pneumatic transport reactor	15–30 MPa 800–1600 °C	Synthesis gas (CO and H_2), energy	Co-gasification with municipal waste, coal, heavy oil
Hydrogenation	Bubble column	20 MPa, 500 °C	Syncrude Bitumen	25 % mixture with vacuum residuals, capacity approximately 40 000 tonnes year^{-1}
Pyrolysis (thermolysis)	Fluidised bed, rotary kiln, tubular crackers	400–900 °C	Wax, oil, gas, energy	Mainly pilot plants, with capacity of e.g. 15 000 tonnes year^{-1}
Reduction in a blast furnace	Blast furnace	2000 °C	Pig iron, furnace gas	

We will now look at a few examples of each process and discuss their important features.

Gasification

Gasification can be defined as the partial oxidation of hydrocarbons in the presence of lower oxygen levels than are required for complete stoichiometric combustion. The main products of gasification are synthesis gases such as CO and H_2. This is already a well-developed industrial process for the gasification of coal and heavy oil fractions, which can be modified for use with plastics waste. The process is carried out at temperatures from 800 °C up to 1600 °C and pressures of 15–30 MPa. Air, oxygen, steam, flue gas, carbon dioxide

Key Facts
● Chemical recycling can (generally) accept mixed or soiled waste and convert them to gaseous fuel or to feedstock for chemical plant.
● Gasification is the partial oxidation of hydrocarbons at reduced levels of oxygen and the main products are synthesis gases such as CO and H_2.

and sometimes hydrogen can all be used as the gasifying agents, either separately or in combination with each other.

Plastics waste is converted into brickettes prior to recycling in fixed bed gasifiers and used as secondary input to supplement coal. There are two general types of gasification technology: fixed-bed and fluidised-bed processes. In both processes, the plastics waste is mixed with coal (lignite) and fed to the gasifier. Plastics and lignite are partially oxidised to synthesis gas $(CO + H_2)$, which can be used as raw material for methanol synthesis or to generate electricity in a power plant. In the process developed, for example, by SVZ Schwarze Pumpe GmbH, gasification is carried out between 800 and 1300 °C at a pressure of 25 bar [9,10]. The process can accept up to 50 % plastic waste.

In the case of fluidised-bed gasification, the mixed feedstock enters the fluidised-bed, dry, at a pressure of about 30 MPa. The products of the process are synthesis gas, which is filtered to remove dust and cooled, to extract the process heat. An example of a coal gasification process in current use is the high temperature Winkler process, where gasification is conducted using an autothermal (i.e. self-heating from the heat of reaction) fluidised-bed reactor at 1000 °C, in an air and steam mixture. Once again, up to 50 % plastics can be admixed with the coal but the quality of the synthesis gas produced can vary depending on the feed ratios of the reactants, temperature and reactant purity.

Fluidised-bed technology can be used not only in gasification but also in other chemical recycling processes, as discussed later, so a more detailed, general description of fluidised beds is given in Textbox 4.1.

Textbox 4.1 Fluidised-bed technology

If a gas is passed upward through a bed of solids with a velocity high enough for the particles to separate and become freely supported in the fluid, the bed is said to be 'fluidised'. Fluidised beds are used in the chemical industry because of their advantages over fixed-beds, including a more intimate contact between solids and gas, the high rates of heat transfer and the uniform temperatures within the bed.

In the applications for polymer recycling, particles of plastic are rapidly melted and coated on to sand particles and thus dispersed throughout the bed. The polymer cracks to lower molecular weight material at the high temperatures of the bed. The higher volatility components vaporise and are collected at the top of the reactor (as syngas, for example), leaving heavier fractions (such as wax) and metallic and mineral fractions to be collected at the bottom. Typically 80–90 % of the waste plastic feed is recovered in one form or the other.

Like other processes, plastics recycling in a fluidised-bed can generate unwanted byproducts. For example, if PVC is not separated out prior to treatment, the principle impurity in the hot gas stream is HCl from the thermal degradation of PVC. There is also a potential for generation of dioxins during this process and other impurities, including volatile metal chlorides. The HCl can be removed efficiently and cheaply by introducing calcium oxide (lime) into the bed. This also makes the fluidised-bed more attractive than a fixed-bed, where the gas clean-up has to be carried out separately. Generation of dioxins can be prevented by operating with a minimum gas residence time of 2 s and maintaining a uniform temperature in the bed, which is relatively easy to achieve in a fluidised-bed.

There are also a number of processes that have been developed specifically for gasification[11] of plastic waste and we will look at just two, namely the Purox system and the Andco-Torrax system, both of which use fixed-bed reactors.

The Purox System[11]

The reactor for this system, shown in Figure 4.6, was developed by Union Carbide and uses pure oxygen as the gasifying agent. Metals are removed from the plastic waste by a

Key Facts

- Fluidised beds are generated by passing a high velocity gas through a bed of solid particles (e.g. sand) which become separated and fluidised in the gas.
- Polymers are gasified in fluidised beds by first melting and coating on to sand particles and then by cracking at high temperatures.
- HCl from the decomposition of PVC can be removed by the addition of calcium oxide.

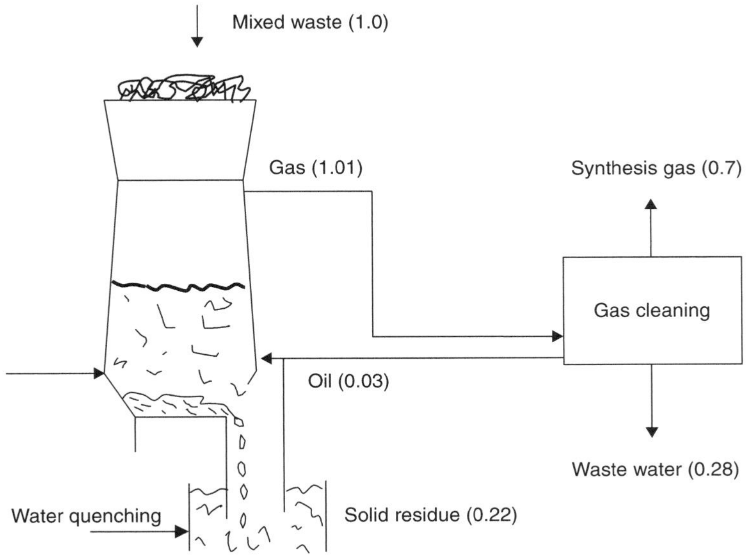

Mixed waste (1.0)

Gas (1.01)

Synthesis gas (0.7)

Gas cleaning

Oil (0.03)

Waste water (0.28)

Water quenching

Solid residue (0.22)

Figure 4.6 The Purox fluidised-bed gasification system[11] (the numbers in brackets in the figure show materials flows based on 1 tonne of solid waste). Reproduced with permission from Xanthos and Leidner, *J. Thermolytic Process, Makromolekulare Chemie (Die)*. Macromolecular Symposia, 1998, **135**, pp. 407–423. Copyright (1998) Wiley-VCH, Weinheim

Key Facts ✂ ❤
● The synthesis gas from the Purox process has higher energy (and economic) value than the nitrogen-diluted gas from the Andco-Torrax process, but the latter exits with a higher temperature, suitable for generating hot water or steam.
● The production of pure oxygen gas is an economically and environmentally costly process.

magnetic shredder before it is fed into the reactor from the top. This allows the waste to dry and then burn more efficiently as it falls into the reactor at temperatures up to 1700 °C. The process generates fuel gas, a mixture of CO, CO_2, H_2 and water vapour. It contains 80 % of the energy of the plastic waste and can be used directly as a fuel, or to produce methanol. However, prior to use it is first cleaned of suspended oils in an electrostatic precipitator and water vapour is removed by condensation. The oil is returned to the pyrolyser and the water is further treated to remove organics. The solid residue remaining in the pyrolyser (about 3 % of the initial waste) is sterilised by the high temperatures, which reduces the disposal costs, but the cleaning unit and the need for pure oxygen make the whole process expensive.

Andco-Torrax System[11]

In this process, the waste polymer is fed straight into the top of the vertical reactor (see Figure 4.7). It dries and burns as it falls in a stream of pre-heated air. Synthesis gas generated during thermal degradation mixes with nitrogen from the air and continues up the bed, where it is then dried before exiting the reactor. Compacted waste in the top zone acts as a seal that prevents the gases from escaping. Oil and wax, also products of degradation, fall into the combustion at the bottom of the reactor and burn, thus maintaining the reactor temperature at between 400 and 550 °C. Some of the oil droplets escape into the gas stream but are scrubbed by the descending waste and returned to the combustion zone.

Dilution of the gas stream with nitrogen means that its calorific value is half that of gas generated by the Purox reactor. However, the lack of a separate gas cleaning unit means that the exit temperature is high (400–550 °C), which makes it ideal for producing hot water and steam for heating.

In summary, the choice of the most suitable gasification process will depend on the composition of the waste stream and the desired

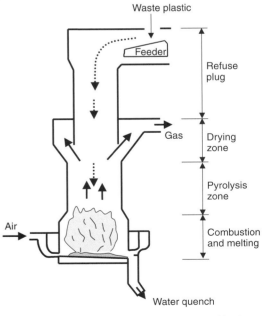

Figure 4.7 The Andco-Torrax fluidised-bed gasification system[11]. Reproduced with permission from Xanthos and Leidner, *J. Thermolytic Process, Makromolekulare Chemie (Die)*. Macromolecular Symposia, 1998, **135**, pp. 407–423. Copyright (1998) Wiley-VCH, Weinheim

output. The highest value product is an H_2- and CO-rich synthesis gas, which can be used in the synthesis of methanol. The lowest value product is nitrogen-diluted fuel gas, which can be used for heat generation. Economically, the production of synthesis gas is obviously preferable, but relies on the availability of a suitably pure waste polymer feedstock and expensive oxygen feed. Gasification has a number of advantages compared to other chemical recycling processes, because of low capital cost and high product value, but it does rely on additional pretreatment processes to separate out waste plastics, which increases the running costs[11].

Pyrolysis

Pyrolysis is thermal decomposition at temperatures from 350–700 °C in the absence of oxygen and other gasifying gases. The polymers decompose to their monomers, oligomers and other organic substances that can be collected separately and used as a feedstock or for energy generation. Pyrolysis is most suitable for waste with high hydrocarbon contents, such as Municipal Solid Waste or mixed textiles waste. However, as in other chemical recycling processes, high PVC contents limit the application of some processes and additional pretreatment is necessary to reduce the PVC content. Pyrolysis of PVC itself yields hydrochloric acid or chloride salts, depending on the presence or absence of hydrogen and metal impurities and, potentially, toxic dioxins.

As there is no oxygen in the system, pyrolysis is not a combustion process but a set of complex reactions, which depend on the type of plastic and the nature of the process used. Several reaction pathways can be defined[11]:

- decomposition into monomers;
- fragmentation of the principal chains into organic components of variable size;
- simultaneous decomposition and fragmentation to monomers/oligomers;
- elimination of simple inorganic components leaving charred residues;
- elimination of side chains, producing complex, cross-linked polymer structures.

The pyrolysis process can be carried out in a variety of reactor systems and we will again look at a few specific model systems.

BP Chemicals Pyrolysis Process

This process is still being developed and is currently operating as a pilot plant in Grangemouth, UK. It was developed and operated jointly by BP, Fina, DSM, Enichen and llochern together with the University of Hamburg[12].

Key Facts ✎⚷
- Pyrolysis involves the high temperature decomposition of polymers in the absence of oxygen.
- Typical products of pyrolysis are monomers, oligomers or others organic compounds.
- Pyrolysis is best suited to waste with a high hydrocarbon and low PVC content.

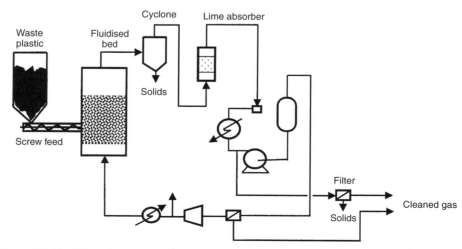

Figure 4.8 The BP pyrolysis process[12]. Reproduced with permission from Xanthos and Leidner, *J. Thermolytic Process, Makromolekulare Chemie (Die)*. Macromolecular Symposia, 1998, **135**, pp. 407–423. Copyright (1998) Wiley-VCH, Weinheim

The BP process[12], shown in Figure 4.8, can be used at a relatively small scale to prepare feedstocks for existing refinery and petrochemical plants. After some preliminary preparation, such as size reduction and removal of most of the nonpolymeric materials, the waste is fed into a heated, fluidised-bed reactor. The reactor operates at 500 °C in the absence of oxygen and produces hydrocarbons, which vaporise and leave the bed with the fluidising gas. This then passes through a hot cyclone to remove solids, such as metals and coke, followed by hot lime to absorb acids and a filter to take out fine solids. The purified gas is cooled to condense out the heavier hydrocarbons, which can then be used in downstream processes, such as production of liquified petroleum gas (LPG) and gasoline products. The remaining, lighter hydrocarbons can be re-used in the reactor as fluidising gas, or as fuel for the reactor.

BASF Thermolysis Process

The BASF process[9], which has a capacity for about 15 000 tonnes annually of plastic polymer waste, integrates both product recycling and energy recovery. The first step in the process, as shown in Figure 4.9 is liquefaction of waste at a temperature between 300 and 350 °C in a cascade of stirred-tank reactors. Pyrolysis takes place in a tubular cracking reactor at temperatures between 400 and 450 °C, followed by a two-step, cooling fractionation, first at temperatures of 330–380 °C and then at 110 °C, to produce mainly naphtha, olefins, aromatics and a heavy fraction. The naphtha can be used as a feedstock to a steam cracker to produce ethylene and propylene (and later polymers); the olefins can be processed to produce alcohol, amines and surfactants, and the heavy fraction can be gasified to produce synthesis gas for the manufacture of methanol.

Key Facts

● Pyrolysis products such as naphtha are steam cracked to form ethylene and propylene.
● Olefins are converted to alcohols and amines and heavy fractions are further gasified to produce synthesis gas, all of which products can be used as feedstocks for chemical synthesis of new materials.

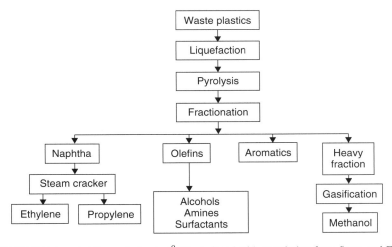

Figure 4.9 BASF integrated thermolysis plant[9]. Reproduced with permission from Sasse and Emig, 'Review: Chemical Recycling of Polymer Materials.' In *Chemical Engineering Technology* (1998), Vol. **21**(10), pp. 777–789. Copyright (1998) Wiley-VCH, Weinheim

Fuji Process

The Fuji process is similar to the two above and is currently operated by Fuji Recycle Industries KK, Japan who have two plants with capacities of 400 and 5000 tonnes annually[9]. The special feature of the process is that ZSM5 zeolites are used to convert up to 80 % of the hydrocarbon gas produced into mixed liquid petrochemical products.

Circulating Fluidised-Bed Pyrolysis System

This system produces the same gases as the Purox System described above, but without the use of oxygen[11]. It uses two circulating fluidised beds with sand as the fluidising and heat transfer medium; one bed is used for pyrolysis at temperatures between 800 and 850 °C

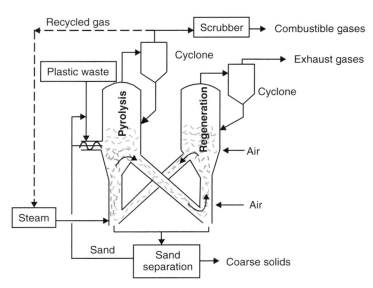

Figure 4.10 Circulating fluidised-bed pyrolysis system[9]. Reproduced with permission from Sasse and Emig, 'Review: Chemical Recycling of Polymer Materials.' In *Chemical Engineering Technology* (1998), Vol. **21**(10), pp. 777–789. Copyright (1998) Wiley-VCH, Weinheim

Key Facts ▬━━◀◗
● The capital cost of a hydrogenation plant is high, but this is offset by the high value of the synthetic crude oil product.

and one for regeneration at 950 °C (see Figure 4.10). Solid waste, fed into the pyrolysis bed, is fluidised by superheated steam. Some of the products of the pyrolysis process are removed from the reactor to generate steam and some carbonaceous products are burned in the regeneration bed to generate heat for the pyrolysis process.

Hydrogenation

Catalytic hydrogenation was originally developed to gasify coal as methane and for the treatment of vacuum residues that are produced in large amounts by the petroleum refining industries. Plastic waste can also be introduced into a hydrogenation unit to produce bitumen and a synthetic crude oil, known as syncrude. In order to allow the waste to be mixed to higher proportions with vacuum residues, it is first depolymerised in a separate reactor. The melt from the polymerisation reactor is mixed with the residues and converted into syncrude by reaction with hydrogen. The syncrude is further refined for use in the petrochemical industry. Although hydrogenation processes have high capital costs, they are quickly recovered through the high value of the products.

The Bottrop and Bergius-Pier processes are two typical examples of hydrogenation technology.

The Bottrop Process[13], shown in Figure 4.11, was developed by Veba-Oel AG. The plastic feed is depolymerised at a temperature of about 420 °C; the hydrocracking process then occurs in a bubble-column reactor in hydrogen at a temperature of 480 °C and pressure of 20 MPa. The main gaseous products are hydrocarbons and ammonia and the solid products are bitumen and syncrude. The main problem is that the process is sensitive to the presence of heteroatoms (*e.g.* sulphur, chlorine, nitrogen, *etc.*) in the polymer, which limits its usability.

The Bergius-Pier process[13] is similar to the Bottrop process, but has the advantage of being able to handle heteroatoms efficiently by binding them to hydrogen to form acids, which are then neutralised to an easily deposited salt.

In summary, chemical recycling is a form of material recycling, which is particularly well suited to mixed plastics waste. These technologies, many still under development, break the plastics down into their chemical constituents that can be used to manufacture a wide range of new industrial intermediate and consumer products. In effect, the plastics are reprocessed at the place of origin, the petrochemical complex. This can be compared to

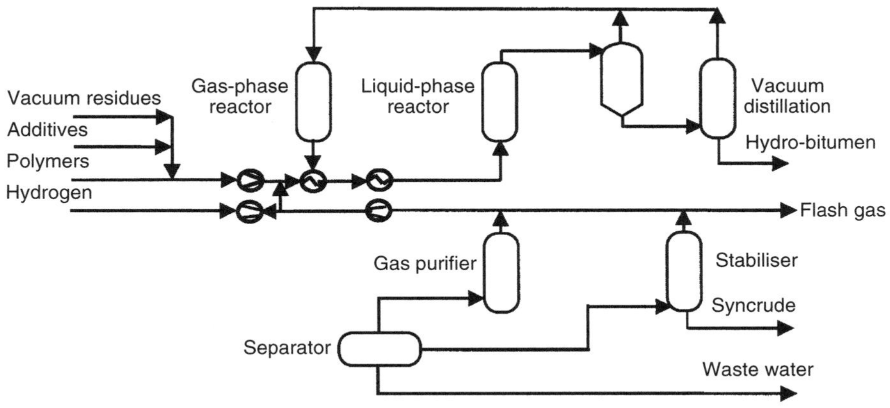

Figure 4.11 Flowchart of the Bottrop hydrogenation process[13]. Reproduced with permission from Mackey (1995), Review of Advanced Recycling Technology. Chapter 14 in '*Plastics, Rubber and Paper Recycling.*' ACS Symposium Series. Vol. **609**, pp. 161–169. Edited by Rader *et al*. Copyright (1995) ACS

paper recycling, whereby the waste paper is converted back to pulp for reprocessing into new paper-based products.

Many of these processes require high capital investment, which means that an additional 'gate fee' will be required in order to compensate for the investment costs of the unit. However, the payback period is reduced by the usually high value of the recovered products. In many cases, existing oil refinery and petrochemical plants can be adapted for plastics recycling, which reduces the capital costs.

The use of plastics as a chemical reactant in the production of the steel is another means of chemical recycling, but because it uses existing facilities it does not require capital investment. This option is described below.

Blast Furnace Reduction

Iron ore is reduced in a blast furnace using reducing agents such as carbon, carbon monoxide or hydrogen, and polymer waste agglomerate can be used as a substitute for heavy oil. Polymeric material is blown into the bottom of the blast furnace at a temperature of 2000 °C where it pyrolyses to form reducing gases and, at the same time, provides a source of heat. Hence, this process spans both chemical recycling and energy recovery. Because the chemical properties of polymeric materials and heavy oil are similar, heavy oil can be substituted by the same quantity of polymer. Almost 80 % of the gases generated are utilised through a long blast furnace moving bed. As with most pyrolytic processes, the chlorine content must be kept low (<2 %) to avoid chlorination and acid formation.

As is the case with other recycling technologies, some treatment of the plastics waste is required to ensure the specification of the recycling process is met. Although such treatment is generally less demanding than for example for mechanical recycling, an additional gate fee will be charged by the reprocessors.

Energy Recovery

If material recycling is for some reasons not viable, then energy can be recovered from the polymer waste as it has a high calorific value. Similarly certain polymeric products that have been removed from the waste stream destined for mechanical recycling may be recycled to recover energy, by either:

● burning in a municipal waste incinerator where plastics, together with other waste material, contribute to the generation of energy for heat and electricity;

● co-combustion, or mono-combustion, where plastics replace another fuel in varying proportions, thus displacing the need to use primary fossil fuels (*e.g.* in cement kilns).

Key Facts ✂
● Plastics waste can be substituted for coal as a source of carbon in blast furnaces used for steel production, provided the PVC content is low.
● Plastics have a high calorific value and their stored energy can be recovered by burning them, either in an incinerator, or by substituting them for other fuels.

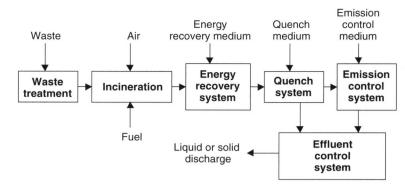

Figure 4.12 Flow diagram of an incineration system generating energy

Key Facts

● Incineration
without energy
recovery reduces
the volume of waste
but is not a
sustainable
recycling option,
because it wastes
valuable
non-renewable
resources and
generates air
emissions;
combustor design
and composition of
the waste are
important factors in
reducing toxic
emissions.

A schematic layout of an incineration system is shown in Figure 4.12. The type of incinerator used will influence the efficiency of energy recovery and emissions from combustion. The latter is a particularly sensitive issue because of public concerns over the possible health and environmental problems. Modern incinerators can keep the emissions to a minimum, provided the right operating conditions are chosen, depending on the waste properties and quantities to be treated. In the following sections we therefore examine the main incinerator types and discuss their advantages and disadvantages with respect to some of these issues. We will then look specifically at some of the issues surrounding emissions, from a scientific and technological viewpoint.

It should be noted that incineration without energy recovery is also possible but is not acceptable from the sustainability point of view because it only reduces the amount of waste, while wasting valuable nonrenewable resources. We do not therefore include it as one of the recycling options in this book.

The main types of incinerator currently in use can be categorised as:

● mechanical stoker,
● rotary kiln,
● fluidised-bed.

The stoker and rotary kiln designs have been in use since the 1970s and have the advantages of being relatively cheap to install and run, whereas the fluidised-bed technology is still under development and more expensive. It does have advantages of efficiency over conventional designs and also of better temperature control and distribution, which is important in controlling emissions.

Mechanical-Stoker Incinerator

This is the main type of incinerator for municipal solid waste. The waste is fed into the combustion zone by the operation of stoker gates or by a simple screw feed (see Figure 4.13) and heat is recovered using an exhaust heat boiler or as electrical power by utilising steam turbines. The schematic in Figure 4.13 shows a typical layout, with an integral boiler, which uses the hot combustion gases to generate steam or hot water.

Rotary Kiln Incinerator

The rotary kiln design[14] is similar to the conventional mechanical stoker design above except that combustion occurs in an inclined, rotating cylinder (Figure 4.14). The pre-sorted plastic waste is fed by the rotating action of the kiln into the combustion zone at temperatures of about 1000 °C. The main advantage of the system is that the percentage of unburned material can be as low as 3 %, but the technology is such that it is expensive and really only suitable for small applications, such as cement manufacture.

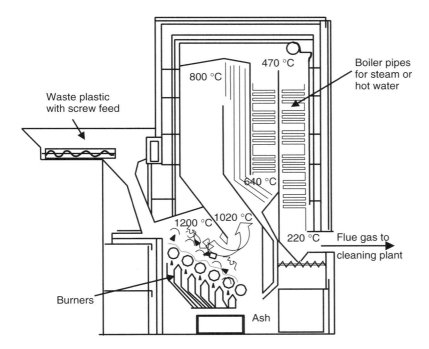

Figure 4.13 Schematic of a conventional mechanical-stoker incinerator

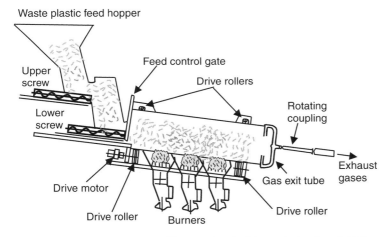

Figure 4.14 Body of a rotary kiln incinerator[14]. Reproduced with permission from J. Kabovoc (ed.). 'Recycling of Polymers'. In *Macromolecular Symposia*, **135**, 1998. Copyright (1998) Wiley-VCH, Weinheim

Fluidised-Bed Incinerators

Modern incinerator types are based on fluidised-bed technology, because of their simplicity of operation and freedom from problems associated with residual unburned fractions of waste. They are also suitable for MSW and direct combustion of waste plastic, rubber and tyres[15].

The fluidised-bed incinerator, shown in Figure 4.15, uses sand as the combustion zone, because it can stand the high temperatures generated by the high calorific values of the waste stream. Incombustible materials such as metal and stones are taken from the bottom of the sand bed after the waste is incinerated. The temperature is more or less uniform throughout the bed. For more detail on fluidised-bed technology, including the emission prevention methods, see Textbox 4.1.

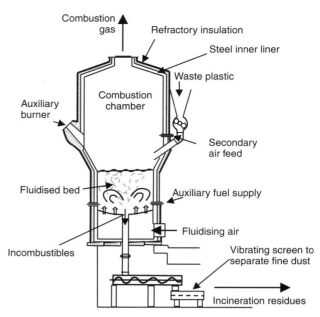

Figure 4.15 Structure of a fluidised-bed incinerator[14]. Reproduced with permission from J. Kabovoc (ed.). 'Recycling of Polymers'. In *Macromolecular Symposia*, **135**, 1998. Copyright (1998) Wiley-VCH, Weinheim

Advantages of this type of incinerator are that:

- the combustion is easy to control;
- the system operation is simple;
- the exhaust gas treatment is not complicated;
- waste volume reduction is very good.

Energy can be recovered in the form of hot water, steam or electricity. Electricity generation is more economic in a larger plant, so small and medium plants tend to go for production of hot water or steam. Plant operating conditions vary widely around the world. The power generating incineration plants in Europe and the United States have steam conditions in a pressure range of 4–10 MPa and temperatures of 370–500 °C. Germany uses pressures in the range up to 20 MPa and temperatures up to 540 °C, but Japan uses temperature and pressure conditions of around 2.5 MPa and 280 °C. Lower temperatures mean lower thermal efficiency, but prevent high temperature corrosion by chlorine, which can reduce generation efficiency by up to 14 %[14]. Designs that superheat the steam from the exhaust heat boiler using the exhaust gas from the power generating gas turbine increase the overall efficiency by as much as 30 %.

New technologies are continually being developed to process waste in all shapes and forms and the reader is referred to Chapter 8 for a summary of some of the newer ideas and processes currently being developed.

Incinerator Emissions

As we have already highlighted, energy recovery from end-of-life plastics can only be considered a sustainable option if the environmental costs of incineration do not exceed the benefit of energy recovery. Hence, it is important to minimise the emissions from incineration. The gaseous pollutants often present in the flue gas include CO, HCl, SO_2, NO_x, particles, heavy metals and dioxins and furans. The latter two often collectively referred to as simply dioxins, are in particular a subject of a continuing debate between the proponents and opponents of incineration over the levels of emissions and the related human toxicity. The literature body on the formation and control of dioxins is vast and a further discussion

Table 4.4 Comparison of the existing and proposed EC Directives on incineration[17]. Reproduced with permission from NSCA (2000). *Pollution Handbook 2000.* Copyright (2000) National Society for Clean Air and Environmental Protection, Brighton.

Pollutant	EC Directive 89/369 (for MSW > 3 tonnes h^{-1}) (mg Nm^{-3})^{a,b}	Proposed Directive (covering both MSW and hazardous waste) (mg Nm^{-3}, as daily average)
Total dust	30	10
Heavy metals (total):	6.2	0.6^c
Pb + Cr + Cu + Mn	5	
Ni + As	1	
Cd + Hg	0.2	
Hydrochloric acid (HCl)	50	10
Hydrofluoric acid (HF)	2	1
Sulphur dioxide (SO$_2$)	300	50
Carbon monoxide (CO)	100	50
Organics (as total C)	20	10
NO$_x$ (NO and NO$_2$)	–	200
Dioxins and furans	–	0.1 ng m^{-3d}

^{a}Emission limits are expressed as a function of the capacity of the incineration plant; the example shown corresponds to the plant with the capacity of greater than 3 tonnes h^{-1} of waste. Combustion gases must be kept at least 850 °C for 2 s in the presence of at least 6 % oxygen. The emissions are expressed based on a standard gas condition of 10 1. 325 kPa, 0 °C, 11 % oxygen or 9 % carbon dioxide, dry gas.
^{b}Nm^3 Gas volume at standard conditions (N stands for 'normal' or standard conditions)
^{c}Expressed as average value over sample period, min. 30 min, max. 8 h.
^{d}Expressed as average value over sample period, min. 6 h, max. 8 h.

of this subject is outside the scope of this book. However, for a technically minded reader, a good overview of dioxin formation and control mechanisms can, for example, be found in a paper by Acharya *et al.*[16]

Most countries now regulate gaseous and solid discharges from incineration and, as an example, Table 4.4 lists the existing and proposed EC gaseous emission limits. It is apparent that the proposed emission limits are in some cases much more stringent than the existing values, particularly for heavy metals, HCl and SO$_2$. The proposed directive also includes limits on dioxins, which the existing directive does not regulate.

These new emission limits can be achieved by the use of best available techniques, which should include the following:

- a control system for the supply of primary and secondary air (to control emissions of CO, NO$_x$ and dioxins);
- control of temperature and residence times in the combustion chamber, boiler and flue treatment units (to prevent formation of dioxins);
- a flue gas treatment system, including cyclones, electrostatic precipitators and filters (for fly ash removal), scrubbers (to remove HCl, HF, SO$_2$), activated carbon beds (for heavy metals and dioxins) and selective catalytic reduction (for NO$_x$ and dioxins).

A recent study sponsored by the plastics industry investigated co-combustion of plastics with MSW[18] and emissions from incineration. The study concluded that the addition of plastics waste to MSW can be beneficial to the combustion process. Depending on the quantity added, there can be an improvement in combustion efficiency, together with an overall improvement in combustion stability. According to this study, there was no increase in dioxins concentrations in the effluent gases, even with the inclusion of 8–10 % w/w of PVC (precursor for the formation of dioxins), which is the European market average. With the addition of lime and activated carbon to the scrubbing system, the concentrations of dioxins could be kept to less than 10 % of the permitted daily limit (*i.e.* 0.01 ng m^{-3}). The study also examined the potential for energy recovery of polyurethane- and polystyrene-based building and insulation foams which can present special challenges for waste management

Key Facts
- Incinerators emit potentially toxic products, the levels of which are regulated by national and international standards.
- CO, NO$_x$ and dioxins in effluent gas can be minimised by controlling the operating conditions of the incinerator.
- Dioxin emissions are further limited by controlling combustion temperature and residence times.
- Flue gas scrubbers are used to control ash, acids, heavy metals, NO$_x$ and dioxins.

because of the presence of chlorofluorocarbons (CFCs) used at one time as blowing agents, and of flame retardants based on bromine, chlorine and phosphorus. Trials in this study demonstrated that co-combustion with MSW is an environmentally recommended option for these foams.

4.4 SUMMARY AND LEARNING OUTCOMES

In this chapter we have examined and discussed some of the current technologies available to identify, sort, separate and process polymer waste. After studying the material in this chapter, you should be able to understand the following technological aspects of polymer waste management:

- the main recycling options and when it is appropriate to use them;
- how polymers can be identified in the waste stream and sorted from it;
- the main technologies for mechanical and chemical recycling and energy recovery;
- advantages and disadvantages of different options and their implications for sustainability.

4.5 REFERENCES AND FURTHER READING

1. SAE J1344 (July 1997). www.sae.org, accessed on 17 July 2001.
2. Plastics – Generic identification and marking of plastics products (ISO 11469), http://www.iso.ch/iso/en/Catalogue (2001).
3. Engstrom, K. (1995). Industrial methods of separating and recycling bottles and containers from the public waste stream. In *Recycling and Recovery of Plastics*, eds. Brandrup, J., Bittner, M., Menges, G. and Michaeli, W., Hanser, Munich, pp. 202–215.
4. Michaeli, W. and Breyer, K. (1998). Polymer recycling – status and perspectives, Makromolekulare Chemie (Die), *Macromol. Symp.*, **135**, 83–96.
5. The REVIVE System, Process for the re-use of mixed plastic wastes, Commission of the European Communities, DG XIII/D3, Jean Monnet Building, rue Alcide de Gasperi, L2920 Luxembourg.
6. Drelich, J., Payne, T., Kim, J.H. and Miller, J.D. (1998). Selective froth flotation of PVC from PVC/PET mixtures for the plastic recycling industry, *Polym. Eng. Sci.*, **38**(9), 1379–1387.
7. Forssberg, E., Shent, H. and Pugh, R.J. (1999). A review of plastic waste recycling and the flotation of plastic, *Resour. Conserv. Recycling*, **25**, 85–109.
8. Chen Chong Lin (1998). Recycling technology of poly (ethylene terephthalate) materials, Makromolekulare Chemie (Die), *Macromol. Symp.*, **135**, 129–135.
9. Sasse, F. and Emig, G. (1998). Review: chemical recycling of polymer materials, *Chem. Eng. Technol.*, **21**(10), 777–789.
10. APME (1995). *Life Cycle Analysis of Recycling and Recovery of Households Plastics Waste Packaging Materials. Summary Report*, Association of Plastics Manufacturers in Europe, Brussels.
11. Xanthos, M. and Leidner J. (1998). Thermolytic process, Makromolekulare Chemie (Die), *Macromol. Symp.*, **135**, 407–423.
12. Hardman, S. and Wilson, D.C. (1998). Polymer cracking – new hydrocarbons from old plastics, *Macromol. Symp.*, **135**, 113–120.
13. Mackey, G. (1995). A review of advanced recycling technology. In *Plastics, Rubber and Paper Recycling, ACS Symp. Ser.*, **609**, 161–169.
14. Akehata, T. (1998). Energy recovery, Makromolekulare Chemie (Die), *Macromol. Symp.*, **135**, 359–373.
15. Schnecko, H. (1998). Rubber recycling, *Macromol. Symp.*, **135**, 327–343.
16. Acharya, P., DeCicco, S.G. and Novak, R.G. (1991). Factors that can influence and control emissions of dioxin and furans from hazardous waste incinerators, *J. Air Waste Manag. Assoc.*, **41** (12), 1605–1615.

17. NSCA (2000). *Pollution Handbook 2000*, National Society for Clean Air and Environmental Protection, Brighton.
18. Mark, F.E. and Vehlow (1998). *Co-combustion of end-of-life plastics in MSW combustors, APME's Technical & Environmental Centre Work 1992–98. A Summary Overview*, APME, Brussels.

Further Reading

Alfons, B.G. and Schoeters, J.G. (1998). Technical methods in plastics pyrolysis, Makro-molekulare Chemie (Die), *Macromol. Symp.*, **135**, 6381.

Audiosi, G., Bertini, F., Beltrame, P., Bergamasco, L. and Castelli, A. (1998). New chemical recycling methodologies: hydrous pyrolysis to recover monomers from polyolefins, Makromolekulare Chemie (Die), *Macromol. Symp.*, **135**, 175–182.

Buffi, R. (1999). Options for poly(vinyl chloride) feedstock recycling: results of research project on feedstock recycling processes, *Plastics Rubber Compos.*, **28**(3), 131–135.

Dintcheva, N.T., Jilov, N. and La Mantia, F.P. (1997). Recycling of plastics from packaging, *Polym. Degrad. Stab.*, **57**, 191–203.

Miltz, L. (1998). Approaches to deal with the issue of plastic packages and the environment, Makromolekulare Chemie (Die). *Macromol. Symp.*, **135**, 265–275.

Sakai, M., Hashimoto, R., Kaneko, M., Kobayashi, K., Furutani, T. and Nakamura, T. (1998). Basic studies on gasification characteristics of organic model wastes, *Combust. Sci. Technol.*, **138**, 381–396.

4.6 REVISION EXERCISES

1. List the main recycling options and describe briefly each of them.

2. What processes are available to separate polymers in the waste stream? Describe the principles of separation by selective solubility.

3. Describe the essential elements of a fluidised-bed reaction. How would conditions be varied to achieve (a) gasification, (b) pyrolysis?

4. What sort of products would you expect to see from (a) and (b) above?

5. What techniques are available for mechanical and chemical recycling? Describe one technology for each option in detail.

6. How can waste polymers be used in combustion processes? Describe one process in detail. List the important features required of an incinerator operation to ensure minimum impact on the environment from gaseous emissions.

7. How would you devise an integrated waste management system for the following types of polymer waste:

   (a) polymers mixed with other municipal solid waste;
   (b) mixed polymer waste;
   (c) waste stream containing only PET plastic bottles with HDPE caps;
   (d) mobile phones?

   Consider each case separately.

DRIVERS AND BARRIERS FOR POLYMER RECYCLING: SOCIAL, LEGAL AND ECONOMIC FACTORS

Chapter 5 – Love Leading The Pilgrim (E Burne-Jones, 1896–97)
Chaucer's "Romaunt of the Rose" provided the literary inspiration for this painting. In the present context, the "pilgrim" seeking the goal sustainable development is led through the tangled briars by a mixture of social, legal and economic arguments.

5.1 INTRODUCTION

The fact that polymer recycling is not currently more widely applied cannot be attributed to any single reason. Although more recently various plastic waste management policies have been put in place in Europe and elsewhere (see Chapter 1) to accelerate recycling, a number of other factors still influence and limit recycling. Some of these are related to the difficulties associated with recovery of post-consumer plastics and their separation from other materials. There is also the issue of quality, which may lead to the inferior performance of recycled polymer materials compared to products made from virgin materials. In addition, the economic costs have historically not favoured recycling. The environmental impacts of recycling must also be considered to ensure that they do not outweigh the benefits. Finally, there are a number of social issues, including a general lack of awareness in, and motivation for, the public to recycle. Even those who recognise the importance of sustainable resource management are often confused as to whether recycling is beneficial overall or just generates additional economic costs and environmental impacts. The social unacceptability of waste management facilities, particularly when employing incineration, has also limited plastics recycling. This chapter examines some of the issues that have thus far prevented the widespread acceptance of polymer recycling by both consumers and manufacturers and which may continue to do so in the future. Several examples are discussed to illustrate how these considerations affect current and possibly future recycling applications. This discussion is continued in Chapter 7, to examine in more detail how environmental considerations affect the selection of the best recycling options.

The following sections investigate the factors that influence recycling by dividing the recycling process into two major life cycle stages: recovery and reprocessing of post-consumer polymeric waste and reprocessing of polymers. Each one of these stages has its own determinants (*e.g.* consumer participation and cost of reprocessing, respectively) which can either encourage or discourage recycling. However, there are also common factors that apply to recycling as a whole (*e.g.* policy issues and public acceptability); these are examined towards the end of the chapter. For reasons explained in Chapter 3, we mainly concentrate on post-consumer plastic wastes, which are primarily household waste and waste from industrial and commercial enterprises.

5.2 RECOVERY OF PLASTIC WASTE: LOGISTICS AND SOCIO-ECONOMIC ISSUES

Successful recovery* of any post-consumer waste, and hence polymers, will depend on many factors, including collection logistics, material sorting, costs, and consumer participation. These factors are discussed in more detail below.

5.2.1 Logistics and Sorting

There are different ways to organise collection logistics of post-consumer waste, including various 'bring' options and 'kerb-side' collection schemes. Known as 'reverse logistics', they often require new collection and transport systems. One of the difficulties in recovering waste polymers from users is the variety of different products involved and their dissipated use within a number of sectors, as shown in Figure 5.1[1]. In Europe, Municipal Solid Waste (MSW), which includes waste arising from households, offices, retailers and restaurants, generates the majority (over 60 %) of plastic waste. The distribution and large industry sectors contribute 20 %; the automotive, construction and demolition, and electrical and

Key Facts
- Collection, separation and identification are not the only issues that limit recycling. Other issues are: quality of product, economic costs and lack of public awareness and motivation.
- NIMBY (Not In My Back Yard) inhibits development of incineration and recycling centres.
- Various collection schemes including 'bring' and kerbside' schemes are known as 'reverse logistics', *i.e.* collection rather than distribution.

---

* For the purposes of the discussion in this chapter, the terms 'collection' and 'recovery' are taken to have the same meaning and are used interchangeably.

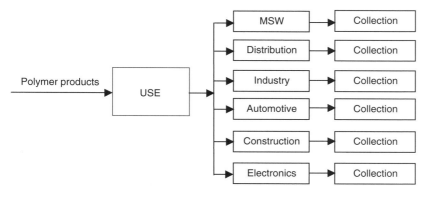

Figure 5.1 Users from whom waste polymers have to be recovered

Key Facts
- 'Bring' and 'kerb-side' schemes are best suited to collection of household waste.
- Mixed kerb-side waste will generally go to energy or fuel recovery; 'dedicated' collections require the householder to recognise and sort plastics, which is not always successful.
- 'Kerb-side' collections are more convenient to consumers and therefore more successful than 'bring' schemes.
- In 1997, the UK had 255 kerb-side collection schemes in operation and 3000 deposit banks for plastics.

electronic products industries contribute 5 % each. The flows of waste plastic materials were discussed in more detail in Chapter 3.

The type of waste and users from which the polymers have to be recovered will influence the choice of collection schemes and transport systems. For example, 'kerb-side' and 'bring' schemes are best suited for household waste. Many communities are already practising kerb-side collection whereby mixed or sorted waste plastic materials are put out generally at the front of the house, frequently in a dedicated container. The waste handlers then collect the waste and take it to the materials recycling facility (MRF). Mixed kerb-side collection schemes are organised in such a way that the homeowner does not have to look for the correct recycle symbol (see Chapter 4 for the symbols). Although this scheme is less demanding on the consumers, as they are not required to sort the waste by type of plastics, it does, however, require post-collection sorting in an MRF, and can limit the recycling options, due to contamination of plastic materials. Kerb-side collection of mixed plastics is, for example, more suitable for energy and perhaps fuel recovery, rather than mechanical recycling. Dedicated kerb-side collection, which requires householders to identify and sort individual plastics, is more efficient and less costly; however, it demands a high participation rate of consumers. Furthermore, it may still require additional sorting in an MRF because sorting at source is not always reliable.

Alternatively, in a 'bring' scheme, consumers deposit plastic waste in the recycling 'banks' provided at convenient locations, in the same way as they do for, for instance, glass bottles and cans. One of the advantages of this scheme is that the established networks for other recyclables could be easily extended to include plastics. Furthermore, from the behavioural point of view, consumers are already accustomed to bringing paper and glass to the recycling 'banks', for instance on their way to a supermarket, so that, in theory, adding plastic waste to their recycling bags should not be a problem. However, experience shows that 'kerb-side' collection schemes are more successful in collecting plastic waste than the bring schemes, reinforcing the importance of consumer convenience for successful recovery of waste materials.

Collaboration between local authorities, waste handlers and consumers is essential for the kerb-side and bring schemes to work. Driven by Agenda 21 (see Chapter 1), which requires Local Authorities (LA) to work towards achieving sustainability, many LAs are setting up various schemes to increase recycling of plastics. For example, in collaboration with Recycling of Used Plastic Containers (RECOUP), funded by voluntary contributions from organisations involved in the plastic bottle supply chain (see Chapter 3), the LAs in the UK have set up kerb-side and bring schemes to collect waste plastic materials from households. In 1997, there were 255 kerb-side collection schemes covering over 8 million households, of which 80 schemes collected plastic bottles. At the same time, there were also over 3000 plastic bottle banks located on over 2000 sites[2]. More detail on the collection schemes that are operating in other countries is given in Chapter 3.

Key Facts

● 'Reverse store
delivery' where
vehicles pick up
waste from stores
and drop it off at
the recycling facility
on their return
journey, are
appropriate for
supermarkets, but
industry requires
bulk collection
services.
● Leasing of goods
(e.g. cars and
photocopiers)
makes for easy
recovery at the end
of life, but for
purchased goods,
exchange schemes
are more efficient.
● In many
countries,
'scavengers' make a
living collecting and
recycling plastics
and other goods.

Mixed and dedicated kerb-side collection is also suitable for commercial users (e.g. offices, light manufacturing, restaurants) and retailers. For example, supermarkets generate large amounts of shrink-wrap (clean waste) which can be collected from the retailers' premises by dedicated vehicles. However, the so-called 'reverse store delivery' may prove to be a more sustainable option, if delivery tracks can be used to pick up the waste and drop it off at a reprocessing facility on their way back. Finally, 'bulk collection' schemes are more suitable for the distribution and large industry sectors, such as the construction and demolition, automotive and electrical and electronic industries.

Various pieces of legislation emerging from the EU, including the Packaging, End-of-Life (EOL) Vehicles and Waste Electronic and Electrical Equipment (WEEE) directives (see Chapter 1) are together providing an important impetus for producers to recover plastic waste and they have been setting up their own collection schemes. The more pro-active companies use leasing and renting as they provide an easy mechanism for return of used goods and this is already being undertaken for cars and photocopiers. However, other, common types of equipment, such as consumer electronics and personal computers, are mainly purchased by consumers, which makes end-of-life collection more difficult. For these products, recovery by bring or bulk-collection schemes is more suitable. For example, some retailers have set up a scheme for return of durable consumer goods whereby consumers exchange their old products (e.g. videos, TV sets, furniture, mattresses, etc.) for new ones. These products are either collected from consumers on delivery of new products or are exchanged directly in the shop. In some cases, the price of a new product is reduced by the value of the used product, to encourage consumers to participate in the exchange schemes. Further discussion on incentives that can encourage consumers to recycle is given in Section 5.2.3.

In addition to the kerb-side and bring schemes, in some developing countries, e.g. India, Bangladesh and Tanzania, collection and sorting of recyclable waste is carried out by the so-called 'scavengers' who retrieve the discarded waste from street bins and landfills and sell it to recyclers. For many scavengers this is the only source of income and in some cases it can exceed the official minimum wage. For example, a study of waste collection by scavenging in Dar es Salaam, Tanzania, has found that 600 scavengers collect around 185 kg day^{-1} of plastic materials[3]. A potential exists for a three-fold increase in the collection of solid waste, which would enable scavengers to earn about 3.3 times the minimum official wage. Thus, in addition to environmental benefits through the recovery of material that would have otherwise been wasted, scavenging also provides employment opportunities to the disadvantaged sections of society making this activity more sustainable. Interestingly, scavenging has become a recent phenomenon in Sweden, even though the recycling rate here is already very high. There is a considerable amount of waste discarded by tourists and people on the move, which provides additional resources for those looking to supplement their income.

A considerable amount of waste is discarded by tourists and people on the move

5.2.2 Costs of Recovery

Recovery of plastic waste is also greatly influenced by the costs of collection, which include logistics (transport) and sorting (labour and sorting equipment) costs. Transporting low-density materials, such as moulded or foamed plastics, is generally not economical and using labour or automated equipment for sorting the waste can be costly. Recovery costs are highly variable and very difficult to calculate. They depend on the structure of the recycling programme and the distances between waste handling companies, MRFs and processors. For example, studies commissioned by the American Plastics Council [4] indicate that the average cost of collection of plastics ranges from $560 – $2120 (~£400 – £1500) per tonne of recovered material. Another estimate[5] quotes a much lower figure of $140 *per* tonne (~£100 *per* tonne) for plastics collection. According to the same author, the sorting costs are equal to $200 *per* tonne (~£140 *per* tonne). A German-based study estimates costs of collection and sorting depending on the recycling technique used[6]. For example, the collection costs range from DM 300 to 450 *per* tonne (~£100–150 *per* tonne) for incineration to DM 900 *per* tonne (~£300 *per* tonne) for pyrolysis and gasification, respectively. This compares with DM 375 *per* tonne (~£125 *per* tonne) if the waste is collected to be landfilled. At DM 700–900 *per* tonne (~£230–300 *per* tonne), separation costs are similar to the costs of collection.

A more recent EU study[7] calculates costs of transport for kerb-side collection of PET bottles from households at 255–305 € *per* tonne of bottles recycled. These costs are lower for bring schemes, averaging 196–242 € *per* tonne for transport from a bottle bank to a sorting plant. However, at 474 € *per* tonne, the study estimates that material sorting is by far the most costly activity associated with PET bottle recycling, exceeding even the costs of reprocessing (332 € *per* tonne). In this study the total cost for recycling 1 tonne of PET bottles is estimated at 508–618 €; this figure includes the revenue from the reprocessed material at 540 € *per* tonne. In economic terms, recycling compares unfavourably with incineration and landfilling, which cost 326–392 and 368–434 € *per* tonne, respectively. However, social considerations, *i.e.* in terms of employment opportunities, favour recycling, which generates on average 17 jobs *per* kerb-side collection, while incineration and landfilling each employ only 1.3 people. Furthermore, although the economics of collection, sorting and transportation may be unfavourable, these activities do have only a marginal environmental impact. For example, the energy consumed is approximately 1000–2000 MJ *per* tonne of recycled plastic, which represents only 3–5 % of the energy savings gained by mechanical recycling[8].

Sorting increases the value of recyclates because it is easier to recycle pure materials, but, as shown above, it involves additional cost, which may make it more difficult for recyclates to compete with the virgin material. Sorting is further complicated by the fact that, in addition to different types of polymers, plastic products also come in different colours. Thus, even if the plastic waste is sorted by the type of resin, if it is not sorted by colour, the recycled material can only be used to manufacture products of dark grey colour[9]. As a result, this type of material has little value. Thus, to maximise the economic value of the recyclate, the plastic waste stream must be sorted by both resin type and colour. Consumers can carry out some pre-sorting of plastics, but, as we have already said, there is a limit to their participation in even less demanding sorting of waste. Hence, most of the sorting has to be carried out in an MRF. This introduces a number of technical problems, which are discussed in more detail in Chapter 4. If sorting is carried out manually, then both the rate and the quality of separation can decrease depending on the variety of the components in the incoming plastic stream. This significantly increases the costs of sorting.

In the future, processes for the separation of polymers will be increasingly automated, reducing to some degree labour charges (but also reducing job opportunities). However, this will lead to an increase in capital costs for the sorting facilities. Furthermore, because the economies of scale demand larger recycling plants to make them less costly and hence more profitable, automated sorting will also require high waste feeds. This implies collection over greater distances thus also leading to higher transportation costs (and environmental impacts). Automatic sorting is further limited by the available technologies and their advantages and disadvantages. Technical limitations to recycling are further elaborated in Section 5.3.1.

The logistics and sorting costs can be borne if the waste products have a high residual value, which can be recovered by recycling. However, for the products with a low residual value at the end of their useful life, such as packaging, the logistics costs associated with their return to a recycling facility may outweigh the benefit from the re-use of the material. To reduce these costs, companies seek to optimise logistics. The main variables that influence the collection logistics and the costs are:

- population density;
- customer and recycling facility locations;

Key Facts

- Estimates of recovery costs vary widely and depend on factors such as collection method, transport distances and sorting methods.
- Typical recovery costs in Europe range from *ca.* 200 € *per* tonne for kerb-side collection to *ca.* 470 € *per* tonne for materials recycling.
- To maximise the value of recyclate, it must be sorted by both resin type and colour.
- Collection logistics must be optimised if the value of the recyclate is to be competitive with that of the virgin material.

- waste material quantities and size;
- collection frequency;
- vehicle fleet size;
- vehicle capacities;
- time and availability of trucks and drivers.

For instance, the costs of transport, comprising the costs of vehicle and driver *per* kilometre travelled, increase almost linearly with the size of truck. In the UK these range from £1 km^{-1} for smaller vehicles (*e.g.* vans) to £1.85 km^{-1} for 38-tonne trucks[10]. One of the ways to achieve minimum costs is to locate waste companies closer to larger quantities of waste. For that reason, some companies are already entering into contracts with neighbouring retailers (*e.g.* supermarkets and restaurants) who control sites where large quantities of plastic waste arise. In addition to economic benefits, minimising distances travelled also reduces the environmental impacts.

A recent study of optimum logistics for recovery of the toner bottles used in photocopiers, concluded that, in order to minimise logistical impacts (both economic and environmental), a deliver–collect scheme should be employed[11]. This requires a servicing engineer to collect empty toner bottles from customers on delivery of new ones and take them to a refilling or recycling depot. The study found that the cost of bottle collection increases linearly with distance and ranges from £0.08 to 0.8 *per* toner bottle (equivalent to £160–1600 *per* tonne) for transporting the load over distances of 100–1000 km. On the other hand, the costs decrease exponentially with increased vehicle payload. Hence, the recovery of bottles is only economical if there are enough bottles to fill the vehicle. Where there are not enough bottles, then mixed payload arrangements with another party accessing a similar customer set can be almost as cost (and environmentally) effective.

5.2.3 Consumer Participation

Consumer participation is one of the most important factors that determine whether recovery of waste materials is successful or not. Many factors influence consumer participation, including environmental awareness, socio-economic status (*i.e.* wealth and education), social conditioning, convenience (*e.g.* effort required and space available for storing waste awaiting collection) and incentives used to encourage consumers to recycle. These are discussed in more detail in the following sections.

Environmental Awareness and Education

Although the environmental awareness of the general public is increasing, consumers are sometimes confused as to the real benefits of recycling. This is not helped by the, often conflicting, information reaching the public *via* the media. However, even scientific studies can sometimes send opposing messages to consumers, which is further exacerbated by the lack of reliable data and scientific uncertainty accompanying many of the environmental issues.

It has been widely recognised that education is crucial for promoting the sustainable development agenda and environmental awareness of the public[12]. Education can be improved by dissemination of information to help raise public awareness, but it is not, on its own, very effective in changing people's behaviour[13-17]. Much more can be achieved by integrating sustainability into primary, secondary and tertiary education, which will ensure that pupils and students will work towards sustainable development in their later roles as professionals or consumers.

Since the Earth Summit in Rio de Janeiro in 1992, various activities have been initiated world-wide aimed at integrating teaching of sustainable development into school and university degree programmes. In the UK, for example, in an attempt to encourage the Further and Higher Education (FHE) institutions to introduce teaching of sustainable development into the curriculum, the (then) Department of the Environment, Transport

Key Facts
- Minimising transport distances improves efficiency and reduces environmental impacts.
- Successful schemes require a high level of consumer participation, which can be improved by education and clear information dissemination.

and Regions (DETR) established in 1998 the Sustainable Education Panel. The aim of the Panel is to consider issues on education for sustainable development and to make practical recommendations for action. In its First Annual Report[18], the Sustainable Development Education (SDE) Panel set a goal that by 2010 all further and higher education institutions should have staff fully trained and competent in sustainable development and should be providing all students with relevant sustainability learning opportunities. The SDE Panel has also developed a sustainable development education specification related particularly to engineering education with the objective to develop and implement sustainability education strategies[19]. The intention is to provide students with an understanding of all the sustainability issues involved (see Chapter 1) as well as raising their awareness of how to work and act in a sustainable fashion[12]. Various science and engineering institutions, including the Royal Society of Chemistry, the Royal Academy of Engineering, and the Institution of Chemical Engineers are supporting the introduction of sustainable development into the science and engineering curriculum in the UK and some universities have already integrated the topic of sustainability into their programmes. The University of Surrey is one such example where the engineering curriculum has been re-designed to incorporate all three dimensions of sustainable development (environment, economy and society) into undergraduate and postgraduate courses. Polymer re-use and recycling is one of the topics addressed to raise students' awareness of what can be done for more sustainable use of resources and is also incorporated into the science curriculum at both undergraduate and postgraduate level.

Some sustainability issues are already being taught in further education (FE) in the UK at GCE Advanced level, including waste management. The Salters' syllabus[20], for example, examines several of the issues that we have discussed in this book. Thus, topics like 'Throw it away or not?' examine the issues of bio- and photo-degradable plastics from a chemical perspective and the incorporation of functional groups or bridging groups that are sensitive to degradation (see Chapter 8 for further discussion). Students are also asked to examine the relative merits of producing degradable plastics or recycling commodity polymers.

Further sustainability learning opportunities can be provided through partnerships among schools, local authorities and companies. For instance, during the 1990s, in the north east of England, the Teesside TEC collaborative learning project (involving the local education authority and local industry) offered a whole series of education liaison activities to local teachers and pupils. While the scope of the programme was wide, one significant benefit (in the present context) was to improve teachers' and pupils' understanding of polymers and the concept of sustainable development.

Other activities might involve scientists and engineers visiting local schools and colleges to deliver presentations, or for teachers, students and pupils to visit the relevant companies to take part in discussions or hands-on activities. For instance, ICI Wilton (based in the north east of England) hosted a 5-day environmental placement for a group of primary and secondary teachers to give them the opportunity to see the operation of a polymer plant and discuss topics such as plastics recycling directly with technologists[21]. Similarly, in 1991 an interactive package was developed for use in the Durham LEA Primary Science Caravan[22], a mobile resource laboratory project financed by local industry, County Durham TEC and Durham Local Education Authority. During the preparation, two primary teachers were seconded to ICI Wilton in order to identify ideas that could be converted into 'caravan-based' activities, some of which targeted packaging and waste and recycling. This and similar types of activity are to be encouraged because they have the real potential to contribute to raising the awareness of sustainable development.

Socio-Economic Status and Convenience

As already mentioned, in addition to environmental awareness, another factor that influences consumer participation in the recovery of waste is socio-economic status. Some studies suggest that wealthier and more educated consumers are more likely to recycle. For

> **Key Facts**
> ● Many professional bodies support efforts to introduce sustainability into education at all levels from primary schools to universities.

It is possible to raise the pupils' awareness of issues such as polymer re-use and recycling within the context of sustainable development through the school curriculum

example, a study of collection of plastic bottles for recycling carried out by RECOUP has found out that areas of above average education and wealth tend to be more productive in plastics collection[2]. This could be explained by a more environmentally aware population that is prepared to take the time and effort to recycle. The same study also found that these areas tend to have more bring sites than less wealthy regions, suggesting that these consumers are prepared to take a less convenient recycling option.

Another issue concerning convenience is the storage space in the households required for sorting and storing waste plastics before it is collected or taken to the recycling point. This may be particularly important for urban areas, where the lack of space may severely limit consumer readiness to recycle. Even in suburban areas, the average house size may not allow for another bin or storage space, in addition to the space already allocated for glass, paper and perhaps other materials being recycled in these households.

Encouraging Consumer Participation

Economic incentives are usually the most successful ways of encouraging consumers to participate in recovery of recyclable materials. One such incentive is a deposit-refund system. This has worked particularly well for packaging, so far mainly for glass bottles and drinks cans. In this system, consumers are charged a deposit, which they can retrieve upon returning the empty beverage containers. Such a scheme has been adopted in some European countries, Canada and the USA to include plastic packaging, mainly PET bottles (see Chapter 3). For example, in some states in the USA, the consumer deposits range from 5 % to 10 % of the price of the beverage[23] and, in Canada, over 20 % of plastic packaging is collected through the deposit-refund scheme.

Another economic incentive can be provided through the weight or volume based disposal fees. In this scheme, households are charged *per* kilogram or m^3 of MSW collected to be disposed in a landfill. This system is in place in the USA and in many European countries, including Germany and Switzerland. However, disposal fees are aimed at all types of wastes and cannot target specific materials, such as polymers. Hence, it is not clear what effect this system can have on recovery of waste plastic, particularly given the low density of the majority of the plastics which would not contribute much to the total weight of the MSW and hence the fee paid by the consumer. The deposit-return schemes can, on the other hand, be targeted to specific components of the waste stream and may therefore be more successful in recovery of plastics. However, their limitation is that they may not be practical for more than a few resins and types of plastic products. Furthermore, countries where this scheme has not already been instituted for other materials, may face strong opposition to its introduction by the public, who would perceive this as an increase in product prices.

Non-economic incentives can also be effective in increasing collection of waste polymers. These for example include displays and certificates given to the communities for participation in recycling programmes. They, however, tend to work better in smaller neighbourhoods and towns where the 'sense of community' acts as a main driver in working towards a 'common cause', in this case protecting the environment. In such neighbourhoods, not caring for the environment may be socially unacceptable, thus encouraging people to recycle.

However, the successful recovery of plastics does not in itself mean that plastic materials will be successfully reprocessed for recycling. Similar to the constraints on recovery, reprocessing will also be influenced by a number of factors, including technical, institutional and economic issues. These are discussed in the following sections.

5.3 REPROCESSING OF PLASTIC WASTE: TECHNICAL, INSTITUTIONAL AND ECONOMIC ISSUES

5.3.1 Technical Issues

Technological Developments

We have discussed recycling technologies in detail in Chapter 4 so that this section gives only a brief overview of technological issues that still limit recycling.

Key Facts
● Areas of above average education and wealth are more successful at recycling.
● Incentives such as deposit/refund schemes encourage recycling.

Although technologies for automatic sorting and recycling of plastics are continually developing, many of them are still far from operation at a commercial scale. As already noted in Chapter 4, technologies for mechanical recycling are probably most well established while those for chemical recycling are still emerging. However, mechanical recycling requires sorting of waste while plants for chemical recycling can take mixed polymer waste. One limitation on chemical recycling, however, is that the recycling plant has to be above a certain size to be economical. That requires high (and guaranteed) supplies of feed material, which at present is not achievable due to low recovery rates of waste polymers (as discussed in Section 5.2). For example, further extension of the BASF pyrolysis plant (described in Chapter 4) in Germany had to be stopped, because the minimum amount of polymer waste of 150 000 tonnes year^{-1} could not be guaranteed[24].

Technologies for automatic sorting are also still in development and they pose different challenges. For instance, scanning separation techniques which use X-ray detectors (see Chapter 4) only operate effectively with larger objects and cannot be used on plastics that have already been shredded[9]. Thus, they cannot sort plastics from durable goods (e.g. cars and electronic equipment) because at present the only practical way of recovering materials from this source is by shredding. The main reason for this is that durable goods are not built for ease of dismantling and recycling: they have a number of different materials commingled together, making their separation very difficult. Applying the design for the environment approaches (as described later in Chapter 6) to these products would enable easier separation of different types of materials and additives, including plastics, without the need to shred them first.

The physical separation of polymers currently uses techniques which rely on physical properties of plastics for separation (e.g. density); these techniques are only able to separate polyolefins from other materials, for example, PET from HDPE. More advanced separation techniques, such as froth flotation and supercritical density separation, on the other hand, are not commercially proven yet. However, if these separation and chemical recycling techniques were to become economically viable they would be able to provide clean, high-quality polymers with the same properties as virgin material[9].

Quality of Recycled Plastics

It is desirable for the recyclate to have properties and quality that are essentially close to those of the virgin resin. In addition to sorting limitations, discussed in the preceding section, the quality of plastics is impaired both during their use and in reprocessing. During use, the product is exposed to stress and environmental conditions that reduce its chances of successful reprocessing, as discussed in Chapter 2. Some thermoplastics can be recycled to give a polymer suitable for direct competition with the virgin material (e.g. PET). However, for most polymers mechanical recycling tends to produce a material with poorer properties compared to the virgin material. This means that they will be destined for use in applications with lower performance specification or that the ratio of recycled and virgin polymers will have to be reduced. Depending on the material and application, it is generally accepted that a maximum of 15–30 % of recycled material can be added to the virgin material without seriously affecting its mechanical properties[25]. This amount may have to be reduced for the recyclates that have been through a cascade of uses. For instance, the properties of virgin PET can be retained for up to five reprocessing cycles, after which the flexibility of the material becomes substantially diminished[26]. Other materials deteriorate sooner, as illustrated in Figure 5.2 for the properties of PC, PP, PBT/PC and PP/EPDM (ethylene–co-propylene-co-diene rubber)[25] which are examined after being through the reprocessing loop four times.

The use of recycled material is also limited by the fragmented nature of the recycling business. The lack of specifications and inconsistencies of feedstock, due to this fragmentation make quality control of recycled plastics impossible leading to uncertainty over the properties of recycled plastics. In many cases the specifications demanded by

Key Facts
- Many of the technologies for sorting and recycling are still not commercially developed.
- Rates of recovery of plastics are not yet high enough to support high throughput automated recycling plants.
- Many plastics deteriorate in use and during recycling, such that recyclate cannot always compete with virgin material and must be used to produce lower quality goods.

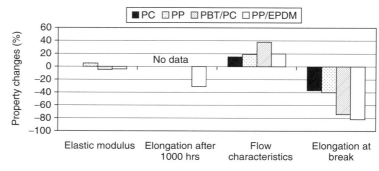

Figure 5.2 Influence of repeated processing for injection moulding of various plastics[25]. Reproduced with permission from J. Kabovoc (ed.). 'Recycling of Polymers'. In *Macromolecular Symposia*, **135**, 1998. Copyright (1998) Wiley-VCH, Weinheim

Key Facts
● Leaching additives from plastics reduces their re-use potential for and poses problems for disposal of contaminated residues.
● Use of re-cycled products is limited by perceptions of inferior quality.

customers are too complex and requirements for some technical materials are too high to risk compromising supplies by the use of recyclates of uncertain quality.

Another barrier to recycling related to quality is the presence of additives, used to modify properties such as colour, thermal stability or processability of polymers (see Chapter 2). The existence of a wide variety of different additives (see Textbox 5.1), some of which are toxic and environmentally damaging, makes it much more difficult to recycle the polymers. For example, most of the pigments and stabilisers (heavy metals) end up in the residues during chemical recycling, making the disposal process more expensive as a result of the increased number of processing operations involved. Many polymers in building materials and cars for instance contain flame retardants with chlorine, bromine, phosphorus or antimony trioxide, which pose environmental, health and safety threats.

Textbox 5.1 Typical additives used in polymers

Anti-oxidants	Flame retardants	Pigments
Anti-static agents	Impact modifiers	Plasticisers
Bio-stabilisers	Light stabilisers	Processing aids
Blowing agents	Metal deactivators	Reinforcing fibres
Compatibilisers	Nucleating agents	Reinforcing and other fillers
Cross-linking agents	Lubricants	Stabilisers

5.3.2 Institutional and Structural Issues

Structural and institutional factors also pose difficulties in the development of polymer recycling[9]. One of the institutional constraints is the perceived inferior quality of recycled polymer, which in addition to the actual quality problems can also limit the use of recyclate. Manufacturers and designers are often reluctant to use recycled polymeric materials because they believe that they may impair the performance of their products or because of their concerns over the health and safety issues[27]. Even if the recycled polymer is significantly cheaper than the virgin material (between 40% and 80%), some manufacturers are still unwilling to use the recyclate[26]. These views may change with the development of quality standards, improved quality control (*e.g.* by using near infrared spectroscopy as discussed in Chapter 3) and a better integration of the recycling market.

The demand for recycled polymers may further be limited by the negative attitudes of companies towards re-used and recycled materials, which then governs their purchasing policy. The policies of some institutions and companies explicitly exclude products which contain re-used parts or recycled materials because their perception is that these products are inferior in specification and hence in their performance. Institutional barriers such as these can severely limit development of a market for remanufactured and recycled products.

However, not all companies behave like this. In fact, one of the important drivers for recycling is the desire of companies to project a 'green image' and so gain a competitive advantage and improve their public relations. Examples include packaging, automobile and office equipment manufacturers who are starting to reclaim the used polymeric (and other) parts and remanufacture or recycle them into new products. One of the now classical examples is Xerox who have been remanufacturing photocopiers for many years by reclaiming old machines from customers and either re-using parts or recycling materials. They are now also investigating possibilities for recycling of other products and packaging, including toner bottles used in photocopiers[11].

Another constraint which acts as a structural barrier to recycling is that polymer manufacturers specialise in one or two types of polymer only[9]. Expansion of polymer recycling, however, requires processing of a number of polymers and that, on the other hand, demands a broader technical expertise. This would require major structural change and capital investment and many manufacturers are simply not prepared to take that route.

As already noted, the recycling business is quite fragmented and that also limits further expansion of the market. At present, there are many potential players in the supply chain from the waste collectors through processors to the product manufacturers and in many cases there is poor integration between them. Attempts to forge partnerships between these players have failed in the past, particularly between large companies (usually manufacturers) and smaller enterprises (usually collectors and reprocessors). This is further explored later in this chapter (Section 5.4.2) in the case study of the EC Packaging Directive. One of the ways to overcome the fragmentation is vertical integration, whereby a company becomes involved with all aspects of recycling, starting from waste collection to the manufacture and marketing of recycled products. In many cases this also results in reduced costs of recycling because of minimised costs of transport, bailing and unbailing scrap, processing and marketing. For example, Milgrom[28] estimates that integration can produce savings between $100 and $220 *per* tonne through lower processing costs and up to an additional $44 *per* tonne by reducing the costs of transportation. Vertical integration is an important factor in determining the long-term economic viability of the polymer recycling industry and is one of the few factors that is under direct control of the industry[9]. Economic aspects of recycling are discussed in the following section.

Key Facts
● Companies looking for a 'green image' actively encourage recycling.
● The fragmented nature of the recycling business inhibits further expansion.
● High capital costs and the volatile nature of the markets for recycled polymers also inhibit recycling.
● Materials recycling offers the most commercially attractive route because of the high value of the products.

5.3.3 Costs of Reprocessing and Market Forces

As we have seen so far, major reasons for the lack of adoption of polymer recycling schemes are poor recovery rates (and hence unreliable supply of waste feeds) and unfavourable economics of transportation. In addition to these, there are also costs of the recycling process itself, requiring energy and materials for its operation. Add to this the often prohibitively high capital costs and volatile markets for recycled polymers, and it is understandable why recycling is not more widely practised. Large companies have found it particularly difficult to offer recycled polymers and to develop markets for these products that are sufficiently large to give them adequate return on capital. Smaller companies which are involved with all phases of the recycling process (from collection to manufacture of finished products) are, on the other hand, more flexible in adapting to the market forces and have so far been more successful in recycling than their larger competitors. Alternatively, companies which are involved in recovery of a number of materials (*e.g.* paper, glass, aluminium and steel cans) can cover the costs of collecting and processing of less profitable materials, such as polymers, from the profits made from more valuable components.

Regarding the recycling process itself, mechanical recycling appears to offer economically the most attractive route for the recovery with the high cost engineering thermoplastics offering the greatest financial benefit[29]. Total costs of mechanical recycling facilities are dominated by the operating costs, mainly because of energy consumption for recycling. Capital costs tend to be moderate, particularly in small to medium size plants. Both of these costs will for any one plant be more or less constant, so that the companies can minimise their overall costs by optimising logistics and sorting costs. Some studies carried out in the USA in 1993 suggested that mechanical recycling of HDPE and PET bottles into flakes or pellets could cost $200–500 *per* tonne of polymer[9,30]. When the collection and sorting costs are included, the recycling of HDPE and PET bottles cost $600–800 *per* tonne and $800–900 *per* tonne, respectively. At that time, PET could sell for $800–950 *per* tonne, which meant that the recyclers could still make a (marginal) profit on the recyclate. The HDPE price, however, was only $600–800 *per* tonne, which in the best case enabled the company to break even and was hence not economically an attractive option. During the following year (1994), the prices of these two polymers dropped dramatically to below the minimum processing costs of around $220 *per* tonne (see Figure 5.3). By 1995, an expanding market for recycled polymers, combined with supply shortages of virgin materials pushed the prices of both PET and HDPE back to $700 *per* tonne and $500 *per* tonne, respectively, in some areas in the USA[31] and recycling became profitable again. However, in 1996 the profitability was once again eliminated, due to new recycling plants coming on stream pushing the prices of recyclate (and virgin polymer) down again. This continued fluctuation of prices of recycled polymers, influenced by so many different factors, makes the market extremely volatile thus discouraging further investment in recycling.

Key Facts
● Many factors affect the profitability of recycling plant and circumstances arise which make recycling not economically viable.

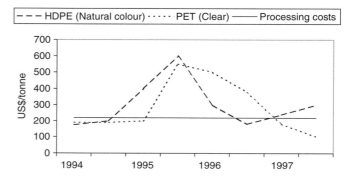

Figure 5.3 Market prices of the major recycled polymers and minimum reprocessing costs. Reproduced with permission from Hadjilambrinos (1996). 'A review of plastics recycling in the USA with policy recommendations.' In *Environmental Conservation*, Vol. **23**(4), pp. 298–306. Copyright (1996) Cambridge University Press

Another study carried out in the USA in 1995 by Mackey[5] focused on chemical recycling by pyrolysis and gasification. For pyrolysis, the costs of feed preparation and processing were estimated at $160 and $200 *per* tonne respectively (see Table 5.1). Adding the costs of collection and sorting of $340 *per* tonne, the total cost of recycling by pyrolysis amounted

Table 5.1 Economic costs of recycling by pyrolysis and gasification[5]. Reproduced with permission from MacKey. 'Review of advanced recycling technology.' Chapter 14 in '*Plastics, Rubber and Paper Recycling*.' (1995). ACS Symposium Series. Vol. **609**, pp. 161–169. Edited by Rader *et al.* Copyright (1995) ACS.

Activity	Costs ($ tonne^{-1})	
	Pyrolysis	Gasification
Collection	140	140
Sorting	200	200
Feed preparation	160	160
Processing	220	180
Total costs	720	680
Selling price of recyclate	120	300
Loss	(600)	(380)

to $720 *per* tonne. The collection, sorting and feed preparation costs for gasification were the same as for pyrolysis, but the processing costs were lower at $180 *per* tonne, bringing the total up to $680 *per* tonne of polymer recycled by gasification. The product from pyrolysis (synthetic crude oil) could at that time sell for $120 *per* tonne making a total loss of $600 *per* tonne. Although the gasification product (syngas) could fetch a much higher price of $300 *per* tonne, that was still not enough to make this option economically viable, resulting in a loss of $300 *per* tonne.

In Europe, the APME[27] evaluated economic impacts of different options for waste management of plastic packaging. The scenarios examined included:

1. 100% landfilling;
2. current situation (1998/99): 15% recycling (12% mechanical and 3% feedstock), 15% energy recovery and 70% landfill;
3. 15% recycling (mechanical), 85% energy recovery;
4. 25% recycling (15% mechanical and 10% feedstock), 75% energy recovery;
5. 35% recycling (25% mechanical and 10% feedstock), 65% energy recovery;
6. 50% recycling (35% mechanical and 15% feedstock), 50% energy recovery.

The study considered collection, separation and processing costs as well as the resulting benefits (income) derived from the recycled products. The results showed that total costs of plastic waste management ranged from 0.17 € *per* kg for landfilling (scenario 1) to 0.67 € *per* kg for recycling of 50% of waste (scenario 6). The most significant costs in all scenarios are the costs of waste collection and sorting, ranging from 0.14 € *per* kilogram (scenario 1) to 0.55 € *per* kilogram of plastic waste (scenario 6). Thus, according to this study, the costs tend to increase proportionally with recycling, so that trebling the recycling rates from 15% (scenarios 2 and 3) to 50% (scenario 6) results in a three-fold cost increase. The benefits of recycling on the other hand quickly level out for recycling rates above 15%, with little difference between scenarios 4 to 6. The same trend was also confirmed by the EU based study[7] on packaging recycling, mentioned in section 5.2.2.

The results of the study pointed to a conclusion that scenario 3 was economically most viable option. The study also compared environmental impacts of these options and found that the same option is also best environmentally. These findings would thus lead to a conclusion that this waste management option is sustainable both economically and environmentally. However, this option proposes recycling of the majority of waste by incineration (with energy recovery) which is not socially acceptable in much of Europe and elsewhere and would face strong opposition from the public. Further discussion on this subject can be found in Section 5.3. Thus, this example illustrates yet again the complexity of the issues involved in waste management and the need to trade off different sustainability components if truly sustainable solutions are to be found and implemented.

A German-based study by Brandrup[32] in 1998 also analysed the difference in waste management costs but in this case, depending on the source of packaging waste. The study estimates how much it costs German society to recycle in particular post-consumer plastic packaging (see Figure 5.4). The figure shows that recycling 1 tonne of post-industrial packaging waste costs approximately DM 800 (or DM 0.8 *per* kg). Recycling the same waste from commercial consumers (*e.g.* retailers) costs twice as much; however, in both cases the total recycling costs fall below the average price of virgin material, making this activity (marginally) profitable. The situation is however quite different with post-consumer plastics. More than DM 2 billion are spent to collect approximately 750 ktonnes and to recycle 500 ktonnes of plastic waste, which corresponds to approximately DM 4 *per* kg of waste. Figure 5.4 shows that the market price of the virgin material is between DM 0.8 and DM 1.5 *per* kg, making recycling of post-consumer waste highly unprofitable. To make up the difference in costs, the 'green dot' subsidy system (see Chapter 3) is used whereby a company buys a sales licence for every individual piece of packaging it sells in German shops. This money is then used to pay for waste collection, sorting and recycling of post-consumer packaging. This study also confirms that costs of logistics dominate the total costs of recycling.

Key Facts
● Costs of recycling increase in direct proportion to volume recycled, because the main cost is recovery of waste.
● The most efficient option for recycling both economically and environmentally is 15% mechanical recycling and 85% energy recovery by incineration.
● Incineration is not socially acceptable so optimum economic solutions have to be balanced against social acceptability.

Key Facts 🔑
● Recycling plastic packaging can be marginally profitable but recycling MSW and other plastic waste is highly non-profitable and must be subsidised by systems such as the 'green dot'.
● The profitability of recycling depends on the price of virgin material (and ultimately on the price of crude oil) and on the costs of landfill, which are set to rise.

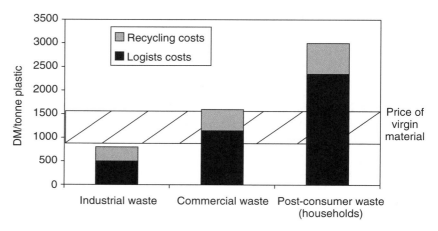

Figure 5.4 Waste management costs depending on the source of waste plastic packaging[32]

As shown in the previous examples, in addition to the costs of recovery and processing, the economic viability of polymer recycling also depends on the cost of virgin material. The price of virgin polymers is very volatile and fluctuations of 50 % or more within a year are not uncommon[9]. The cost of virgin polymers depends on the price of raw materials, *i.e.* crude oil, and on the demand and supply, as determined by the capacity of the manufacturing plants. The price of crude oil influences recycling in two ways.

First, high oil prices push up the cost of virgin polymers making recycled polymers more competitive. For example, a doubling of the price of crude oil from $10 a barrel to $20 a barrel could produce a 20–40 % increase in the costs of ethylene and a 10–15 % increase in the cost of manufacturing PE, which are then reflected in the price of virgin polymer. Although predicting prices of the crude has proved to be a futile activity in the past, most forecasts for the price of crude in the period 2002–2010 range from $25 to $35 *per* barrel. At these prices, polymer recycling may become more attractive so that increasing price of the crude in future may act as one of the important drivers for growth of the recycling business.

Secondly, recycling by pyrolysis, which converts waste polymers into synthetic oil, may be more attractive as crude oil prices rise. In the recent past (since 1973), the cost of crude ranged between $10 and $40 *per* barrel. At $10 a barrel, the most that polymers could be worth as synthetic crude oil would be $80 *per* tonne, which would not be enough to cover the recovery and processing costs. Even at $40 *per* barrel, it could be worth $320 *per* tonne (1993 prices), which would cover the processing costs but probably not the collection costs (see the discussion on pyrolysis above).

Thus, overall, recycling can be considered to be economically viable only if the cost of recycling is equal to or lower than the cost of producing virgin material plus the cost of alternative disposal methods. Given the volatility of the virgin and recycling markets, unless it is subsidised in some way (*e.g.* green dot in Germany), polymer recycling may not be attractive, and other disposal options will have to be considered. Currently, the cheapest economic alternative to recycling is landfilling. According to some USA-based estimates, landfilling costs are around $30 *per* tonne[30]. In the UK these costs are £12 *per* tonne and are set to rise in future. In Germany the landfill cost has already risen from DM 300 to DM 600 *per* tonne between 1990 and 1995. With the escalation of the disposal costs, recycling polymers may become economically more attractive and manufacturers may become increasingly interested in exploiting this market. The plastics industry makes recommendations which, in their view, would increase the competitiveness of plastics and these are listed in Textbox 5.2.

At a forum in May 1999, the Association of Plastic Manufacturers Europe (APME) produced the following general conclusions which would in their view promote recycling:

- local conditions should decide the optimum combination of plastics recovery options in order to achieve the most eco-efficient solution;
- for a number of waste streams, material recycling is, depending on local conditions, the preferred option in environmental terms;
- as the relatively low cost of virgin material is a major factor, it is important to reduce the cost in all stages of the recycling process;
- the focus should be on wastes generated in industry, shops, offices and similar sources of waste as only some specific streams from household waste are of interest;
- a uniform material identification system could be useful for (manual) dismantling of appliances and automobiles, but is generally of little advantage in sorting packaging items;
- minimum requirements for content of recycled materials should only be considered in some specific product/application combinations;
- legislation, which would compromise innovation and improvement of resource efficiency of the use of plastics, should be avoided.

Based on the above comments, the plastics industry recommends the following actions to be taken:

- encourage the maximum diversion of waste plastics from landfill at end of life;
- encourage eco-design, including design for recycling, while ensuring the initial criteria of fitness for use and prevention have been met;
- increase selective collection, dismantling schemes and automatic identification and sorting in order to obtain recyclable homogeneous streams of plastics, at the lowest possible cost;
- encourage product chain responsibility for the development of competitive recycling technologies meeting the market needs (*e.g.* outlets for recyclates, adaptation to new products/new materials which better serve consumers needs);
- develop CEN standards for waste for recycling and for recyclates;
- encourage demand for recyclates particularly in thick walled applications, where performance criteria can be readily met and such use is environmentally sound;
- encourage recycling and the use of recyclates through fiscal incentives, such as reduced rate of value added tax.

Incineration with energy recovery is another alternative to mechanical or chemical recycling. Its costs normally fall between the landfill and mechanical recycling costs. Ellis[30] estimates that incineration in the USA costs around $100 *per* tonne. In Europe, the costs can reach 400 € *per* tonne (~$360 *per* tonne)[7]. However, as discussed in Chapter 4, this option receives strong opposition from the public, due to the fears associated with dioxins and other toxic emissions. Although a properly operated incinerator, particularly with energy recovery, can be far less environmentally damaging than a landfill site, it is still difficult to convince the public and obtain planning permission to build an incinerator. This is particularly the case in the USA, Australia, UK and some other European countries, where only a few of incinerators remain in operation.

As we have already mentioned in Chapter 3, one of the reasons why waste disposal in landfills is still cheaper than recycling is that landfilling is in effect underpriced: it

Key Facts ✂
- A properly operated incinerator is environmentally less damaging than a landfill site.
- Legislation is an important driver for recycling particularly where profitability is marginal.

only takes into account more visible costs, such as waste collection, landfill operation and perhaps closure costs. It does not consider other less tangible costs, such as loss of valuable resources and environmental protection. This underpricing of landfill and associated resources encourages increased consumption and wasteful use of materials and reduces the incentive to recycle. However, placing a price on the 'environment' is not a trivial task and debate on this issue continues. One of the ways to overcome some aspects of this problem is to devise policies that encourage recycling. The following section examines how that might be achieved.

5.4 OTHER FACTORS AFFECTING POLYMER RECYCLING: POLICY ISSUES AND PUBLIC ACCEPTABILITY

5.4.1 Policy and Legislation

An overview of policy and legislation affecting recycling has been given in Chapter 1 so that this section focuses on different policy issues and the influence they have on recycling rather than re-visiting the detail of the legislation itself.

We have seen from the discussion above that recycling of polymers is at best only marginally profitable. Hence, market forces on their own are not enough to provide an incentive for increased recycling. Therefore, other drivers have to be found to encourage recycling practices. Policy and legislation have so far proved to be the strongest impetus for setting up recycling schemes.

The policy options that can encourage recycling include:

- minimum recycled content standards;
- procurement programmes;
- producer responsibility policy;
- integrated product policy;
- subsidies for recycling;
- taxes on the use of virgin material;
- deposit-refund programmes;
- disposal fees.

Minimum recycled content standards specify a minimum amount of recycled material that must be incorporated into products, *e.g.* refuse sacks, packaging containers, *etc.* For example, California requires a 10 % recycled content for plastic refuse sacks while in Wisconsin all rigid containers must contain at least 10 % recycled material[9]. Similar standards can be defined for materials to specify the amount of recyclate that must be mixed with a virgin resin. While product-based standards can be applied in smaller regions (*e.g.* a county or a state), the material-based standards are more effective if implemented in larger areas (*e.g.* within the EU or USA). The reason for this is that materials are often made in one country and then exported to another. Thus, it makes little sense to enforce such a standard in a particular country if the material is not going to be used in that country.

Institutional customers (*e.g.* governments, universities, companies, *etc.*) can devise procurement (purchasing) programmes to promote buying goods with a minimum recycled contents. For example, a government purchasing office can identify their preferential suppliers who incorporate recycled polymers in their products. In the USA, for instance, the federal government obliges all of its agencies to purchase items containing recycled materials if the single item or the quantity of items have a value of $10 000 or more[9]. However, the problem with procurement programmes like these is that they do not target any specific material, so it may happen that they will favour only those materials that are most easily recycled (paper or glass). They are also still relatively rare and hence their impact on recycling, particularly of polymers, has so far been limited.

As we have seen in Chapter 1, producer responsibility is the basis of many EU directives and it has been one of the major drivers for recycling. It shifts the responsibility for the

Key Facts
- Corporate purchasing policies can be developed to target recycled goods; producer responsibility for their end-of-life products is generally the preferred target of legislation and may in future be extended to make producers responsible for environmental impacts of a product throughout its life (Integrated Product Policy).
- Subsidies such as loans or tax credits encourage recycling, but may also encourage the profligate use of material.

recovery from government and requires manufacturers to retain the 'ownership' and recycle their products. The most well known example of one such directive is the Packaging and Packaging Waste Directive (see Chapters 1 and 3), which has been implemented in the EU countries since 1996. The directive stipulates the recovery targets of 50–65 % by weight, of which 25–40 % must be recycled (recovery includes materials recycling, composting and energy recovery). A minimum of 15 % of each material must be recycled as material. A case study discussing issues in implementation of the packaging directive in the UK is discussed later in this chapter. Other EU producer responsibility directives, such as Waste Electronic and Electrical Equipment and End-of-Life Vehicles (see Chapters 1 and 3) also have a potential to contribute towards increased recycling rates. However, they are still to be adopted so that their effect on recycling will not be visible in the near future.

An Integrated Product Policy (IPP) is a more recent policy option to emerge from the EU Commission. This proposed policy would extend the responsibility of manufacturers for the environmental impacts of their product throughout its life cycle. One of the key drivers for the development of IPP is the need to co-ordinate policies throughout a product's life cycle in an integrated manner. So far, policy in the EU (and elsewhere) has been fragmented so it could happen that one piece of legislation would encourage recycling while the other one would act in the exactly opposite direction. One such example is recycling of aluminium. The Climate Change Levy (CCL) introduced in the UK in 2001, which taxes high energy industrial users for the use of fossil fuels and emissions of CO_2, exempts primary aluminium producers from paying the tax. This makes the production of virgin aluminium economically attractive and thus discourages recycling of the material. The latter, however, uses 95 % less energy than the primary production and therefore immediately saves large amount of fossil fuel and CO_2 emissions, which is the primary aim of the CCL. It is hoped that IPP will act to overcome these conundrums.

Subsidies to encourage production of recycled material can be in the form of loans or tax credits. Loans enable businesses to access capital for investment in recycling. This may be particularly effective for small to medium enterprises (SMEs). Tax credits to businesses using recycled polymers in their products enhance the economic competitiveness of these materials and thus encourage recycling[9]. However, some argue that subsidies on their own are not a sustainable policy option because they may reduce the price of recycled polymers and so encourage the extravagant consumption of materials, because they are cheap.

By contrast, taxes on the use of virgin material and deposit-refund programmes (see the section on encouraging consumer participation) may act in the opposite direction and encourage recycling while at the same time discouraging consumption. Taxes on the virgin material, however, are still rare. One example is found in the state of Florida, which imposes a fee for plastics packaging that does not have at least a 25 % recycled content[9]. However, this law faced various political difficulties and remained in effect for only 2 years, a timescale that was too short to have an effect on recycling of materials other than HDPE and PET, which were already being recycled. Deposit and refund schemes are, on the other hand, much better established, particularly in some European countries, as discussed in the section on encouraging consumer participation (and Chapter 3).

Finally, disposal fees may also act to deter from dumping waste plastics and to encourage recycling. These fees have been in existence in Europe for some time and in some countries have risen significantly in the past few years (see Section 5.3.3). However, as already noted, this policy option does not normally target single materials and it is therefore not clear how effective it is for polymer recycling alone.

From the discussion so far it is obvious that no single, isolated policy measure can achieve the desired recycling targets. Instead, an integrated policy with all of the above options combined with market forces and consumer education is likely to be more successful and to lead to more sustainable solutions. This must also be accompanied by the development of appropriate marketing strategies to promote the use of recycled materials in consumer products.

Key Facts
● Disposal fees discourage dumping of waste, but may encourage illegal 'fly-tipping'.
● The development of voluntary agreements on recycling is difficult due to the generally large number of conflicting interests of stakeholders.

5.4.2 Legislation *Versus* Voluntary Agreements: The Case of the Packaging Directive

As we discussed briefly in Chapter 1, the policy options raised in the previous section can be implemented through legislation or by adopting a voluntary approach. Hence, the question often posed in environmental policy is: 'Is legislation better and more effective than voluntary agreements to achieve the policy measures?'. Industry has often argued that legislation constrains development and innovation and should be avoided and that government should instead involve industry in determining how policy will be implemented. However, other stakeholders prefer government to legislate, in order to ensure that the policy will be implemented.

Despite many attempts and interest by industry, there are still few examples of voluntary agreements in use in Europe. A number of barriers can be identified, including the, usually large, number of stakeholders, resources and the time it takes to reach an agreement. The case study of Packaging and Packaging Waste Directive implemented in the UK is discussed below to illustrate why it is often impractical to develop voluntary agreements[33].

The European Commission specifies recovery and recycling targets, but does not prescribe how the Packaging and Packaging Waste directive (see Chapter 1) should be implemented within member states. Thus, each state is free to devise its own policy as to how to achieve these targets. Two options are open to the governments: either to legislate or to leave it to industry to draw up a voluntary agreement, thus negating the need for legislation. If government decides to legislate, then it specifies exactly what must be done (and often how) to achieve the policy as well as the penalties for noncompliance. If, on the other hand, policy implementation is defined through voluntary agreements, then the affected companies decide whether they want to participate and how they are going to achieve the targets. To ensure participation of industry and emphasise that there is no advantage in remaining outside the agreement, an explicit 'threat' (*e.g.* a fine) or incentive (*e.g.* tax reduction) is usually defined. However, for voluntary agreements to work, a high degree of consensus is necessary. If there are many stakeholders in the supply chain, then that may prove a major limiting factor in reaching such agreements.

After the adoption of the directive, the UK government was reluctant to legislate and wanted industry to take a lead in its implementation. One of the reasons for that was that the government believed that that would encourage industry to take a proactive approach, it would cut costs, provide greater flexibility and shorten the time in meeting the targets. However, it was to be proved wrong, as demonstrated by the case study in Textbox 5.3[33]. The latter demonstrates that, much as governments and industry would like voluntary agreements to be achieved and to work, many sectors and issues affected by environmental policy are too complex to achieve the goals using voluntary measures alone[33]. As noted earlier, an integrated approach with combination of policies, either voluntary or legislated, is a much better way forward.

Textbox 5.3 A case study of the implementation of the EC Packaging Directive in the UK[33]

The packaging supply chain is complex and involves a wide range of companies that operate in extremely competitive environments. It comprises four sectors: raw materials suppliers, converters (manufacturers of packaging), packers/fillers and retailers. With such disparate activities within the supply chain, each sector has their own interests to protect within a policy-making process. That created difficulties in deciding who in the supply chain should be responsible for recovery of waste and for collection and distribution of funds. The debate over responsibility for recovery of waste was the greatest stumbling block in the policy-making process, which made reaching an agreement almost impossible.

Valpak Working Representative Advisory Group (V-WRAG), containing members from all parts of the supply chain and set up to negotiate with the government, proposed various options in an attempt to identify a fair, shared responsibility scheme. However, each proposal had failed because different players in the supply chain felt that the others were more favoured and that they would lose out. For example, the material suppliers and converters were trying to shift the responsibility downstream whilst the packers/fillers and retailers were trying to share the obligation so that their share would be reduced.

Realising that the agreement would not be reached by June 1996, as stipulated by the Directive, the government then set up a meeting at the end of 1995 to try and facilitate the process which was by then becoming extremely drawn out. Finally, an agreement was reached which specified a percentage obligation for each player in the chain (see Table 5.2). However, many other issues remained unresolved.

Table 5.2 Allocation of recovery obligation in the UK to achieve the targets required by the EU Directive on Packaging and Packaging Waste[34].

Players in the supply chain	Obligation (%)
Packaging raw material manufacturers	6
Converters	11
Packers/fillers	36
Retailers	47

For example, while the government wanted a single point of obligation, nobody from industry would agree to act, making it more difficult for the government to monitor the implementation. Another complication was related to the fact that businesses may perform different functions in the supply chain so that it would be difficult to predict which sectors would carry the main burden. The prediction was that it would be the large players and the brand owners, *e.g.* Coca-Cola and Schweppes. Furthermore, while the government was reluctant to legislate, industry wanted it to do so because they were worried about 'free-loaders' (*i.e.* those who would not participate in the scheme but would still benefit from it). Ironically, this was the only point on which all players in the supply chain agreed unanimously.

Left without much choice, the government decided in the end to legislate. The required legislation, which was brought into the Environment Act 1995 referring to producer responsibility, states that criminal sanctions may be placed on individual companies if they do not fulfil their obligations. Later revisions followed and the Directive is now intended to be met in the UK through implementation of the Producer Responsibility Obligations (Packaging Waste) Regulations 1997. A few years on, we now know that the Environment Agency has already prosecuted some companies for failing to comply with the Packaging Regulations. The prosecutions have been for various offences including failing to register, failure to meet recycling and recovery targets and failure to submit certificates of compliance.

So, instead of saving time and money, as the government had hoped at the beginning of this process, in the end it had to invest time and resources in the enactment of legislation and the development of regulations. Industry also suffered considerable costs in this process, which was much easier for larger companies than for smaller players in the supply chain. With the benefit of hindsight of course, this money could have been spent in a more efficient way, for instance for setting up recycling schemes, rather than being consumed by drawn out negotiations.

Finally, it should be noted that other stakeholders were excluded from the policy-making process, most notably the LAs, the waste management industry and consumer groups[33]. This prevented exchange of relevant information and made the implementation more difficult because of lack of coordination. However, involving even more organisations in the negotiations would have made the whole process even more complex and lengthy.

Key Facts
● In general, public perception is that 'recycling is a good thing' and saves natural resources, in reality LCA has proved that this is not always the case.
● There is strong public pressure against construction of new incineration/waste management facilities, even though scientific studies have shown their effects on the environment to be far less than those of landfill.

5.4.3 Public Acceptability

Another important aspect that must be considered in devising waste management policies and choosing the most sustainable options is public acceptability: if society is not prepared to accept a proposed option or set of options, then they cannot be considered sustainable. Public opinion about recycling depends on many factors, including historical conditions, cultural background, socio-economic status and personal beliefs.

Historically, recycling of waste has been considered by the general public as a worthwhile activity that contributes towards reduction of waste and reduces the use of primary resources. Prior to the 90s, the benefits of recycling were not questioned much and the public generally assumed that recycling was overall beneficial for the environment. Public pressure which emerged out of these beliefs, has, among other factors, led to setting up of various recycling programmes which are now well established for many materials, including paper, glass and metals. However, with the application of life cycle thinking and Life Cycle Assessment (see Chapter 6) in the early 1990s, it became apparent that some cases of recycling might not be as environmentally beneficial as originally thought. A typical example is that of paper recycling, which in Western Europe is less sustainable than energy recovery from paper waste (see Chapter 6). With the emergence of these findings, the public has started to question the viability of recycling; however, in general they have been slow in changing their attitudes and, for example, still prefer recycling to energy recovery.

One of the main reasons for this is the perceived (and in some cases real) health risks associated with incineration in particular. The principal concerns are trace emissions of toxic compounds such as halogenated dioxins and furans and toxic heavy metals (see Chapter 4). Increased traffic in the area is also often quoted as one of the problems. In addition to these, there is also the 'NIMBY' syndrome (Not In My Back Yard): nobody wants a waste management facility (or any other industrial plant) in the area where they live, although everybody wants the waste to 'go away'. Another argument for rejecting incineration, mainly used by the environmental pressure groups, is that building new incineration plants encourages increased consumption, as people would see the problem of waste as 'solved'. Other reasons for public reluctance to accept incineration include the belief that different technologies for thermal treatment (e.g. pyrolysis and fluidised beds) are more benign and that investment in MSW plants would impede other waste management policies because of the large long-term investment locked-up in these facilities.

Several scientific studies showing that modern incinerators can be environmentally benign and that energy recovery is more sustainable than landfill disposal have so far failed to convince the public to accept this waste management option. The opposition to any new installation of this type is hence very strong and growing in many developed countries in the world, including Europe, Australia and the USA. In Europe, the amount of MSW incinerated has over the past years decreased slightly, including the amount of polymer that is disposed in this way[7]. One of the reasons is that the old incinerators have been closed without being replaced by new installations. This situation is mirrored in the UK, where only 7% of MSW is incinerated[7] and no new installations have been built in the past years, despite many proposals that have been put forward. Australia also has very few incinerators and they are mainly used for hazardous waste (e.g. from hospitals) with only one large-scale installation[35]. A rare exception to this trend is Denmark where 100 % of polymeric waste is incinerated. Sweden and Luxembourg follow close behind with figures of 65 % and 70 % respectively. Japan incinerates 65 % of all MSW and this figure is set to increase. The Japanese position can probably be explained by its large population density and lack of landfill space, which leaves few other options available for dealing with waste.

Public unacceptability of incineration has acted as a serious obstacle to reducing waste volumes disposed in landfills and has constrained informed debate on the issue. This situation has led to the local authorities facing ever-increasing mountains of solid waste and yet not being able to deal with it efficiently. One of the ways to overcome these difficulties is for the LAs and developers to engage in dialogue with the public and try to work with rather than against them. In this way, socially acceptable solutions may be found, thus leading to a more sustainable society.

5.5 SUMMARY AND LEARNING OUTCOMES

This chapter has outlined the major drivers for and barriers to polymer recycling, in both the recovery and reprocessing stages. These have included logistical, technical, economic and social factors. Following on from the previous chapters and upon studying the material in this chapter, you should be able to understand and discuss the following:

- sources and composition of post-consumer polymer waste;
- main waste collection schemes depending on the type of waste;
- logistical and sorting problems and how they can be overcome;
- the factors that influence public participation in recycling;
- the role of education in promoting recycling and sustainable development in general;
- technical and institutional factors that limit recycling and what can be done to improve the situation;
- economic costs of various options;
- the role of policy and voluntary agreements for recycling;
- the importance of public acceptability of waste management options;
- the trade off between different factors and how that may influence the choice of different waste management options.

5.6 REFERENCES AND FURTHER READING

1. APME (1995). *Plastics Consumption and Recovery in Western Europe 1995*, Association of Plastic Manufacturers in Europe (APME), Brussels.
2. Smith, D.N., Harrison, L.M. and Simmons, A.J. (1999). A survey of schemes in the United Kingdom collecting plastic bottles for recycling. *Resour. Conserv. Recycling*, **25**, 17–34.
3. Kaseva, M.E. and Gupta, S.K. (1996). Recycling – An environmentally friendly and income generating activity towards sustainable solid waste management. Case study Dar es Salaam City, Tanzania, *Resour. Conserv. Recycling*, **17**, 299–309.
4. Fisher, M.M. and Liesemer, R.N. (1998). Collection of post-consumer plastics for recycling. A report prepared for the IUPAC Working Party on Recycling of Polymers, American Plastics Council, *Macromol. Symp.*, **135**, 315–325.
5. Mackey, G. (1995). A review of advanced recycling technology. In *Plastics, Rubber and Paper Recycling, Chapter 14, ACS Symp. Ser.*, **609**, 161–169.
6. Bauermeister, U., Maiburg, U., Huber, J., Gutzer, W. and Vick, S. (1994). Wirtschaftlichkeit und stofflich-ökologischer Nutzwert von werkstofflichen, rohstofflichen und energetischen Verfahren der Kunststoffverwertung, *Polymerwerkstoffe '94*, Merserburg, Proceedings, p. 526. Cited in Radusch, H.-J. (1998). Future perspectives and strategies of polymer recycling. Chapter 6: The Way Forward. In *Frontiers in the Science and Technology of Polymer Recycling* ed. Akovali, G., Bernardo, C.A., Leidner, J., Utracki, L.A. and Xanthos, M., Kluwer, Dordrecht, pp. 451–467.

7. PIRA (2001). Evaluation of costs and benefits for the achievement of reuse and recycling targets for the different packaging materials in the frame of the Packaging and Packaging Waste Directive 94/62/EC, Proposed Draft Final Report, RDC Environment and Pira International, May, p. 341.

8. Otto, B. (1997). *Recycling and Recovery of Plastics from Packaging in Domestic Waste: LCA-type Analysis of Different Strategies. Summary document*, Duales System Deutschland AG, Cologne, p. 36.

9. Hadjilambrinos, C. (1996). A review of plastics recycling in the USA with policy recommendations, *Environ. Conserv.*, **23**(4), 298–306.

10. FTA (1998). *Manager's Guide to Distribution Costs July 1998: Section 1 Wages and Section 2 Vehicle Operating Costs*, Freight Transport Association.

11. Mellor, W., Williams, E.A., Stevens, G.C., Clift, R. and Azapagic, A. (2001). *Chain Management of Polymer Materials (CHAMP), Final Report*, University of Surrey, UK.

12. Perdan, S., Azapagic, A. and Clift, R. (2000): Teaching sustainable development to engineering students, *Int. J. Sustainability in Higher Educ.*, **1**(3), 267–279.

13. Boerschig, S. and De Young, R. (1993). Evaluation of selected recycling curricula: educating the green citizen. *J. Environ. Educ.*, **24**(3), 17–23.

14. Hungerford, H. and Volk, T. (1990). Changing learner behaviour through environmental education, *J. Environ. Educ.*, **21**(3), 8–21.

15. Iozzi, L.A. (1989). What research says to educator: part one, *J. Environ. Educ.*, **20**(3), 3–9.

16. Iozzi, L.A. (1989). What research says to educator: part two, *J. Environ. Educ.*, **20**(3), 6–13.

17. Sia, A.P., Hungerford H. and Tomera, A.N. (1985). Selected predictions of responsible environmental behaviour: an analysis, *J. Environ. Educ.*, **17**(2), 31–40.

18. DETR (1999). *Sustainable Development Education Panel: First Annual Report 1998*, HMSO, London.

19. SDE Panel (1999). *Sustainable Development Education: Engineering Specification, Position Paper*, Forum for the Future/DETR, London.

20. Burton, I., Holman, J., Pilling, G., Waddington, D. (1994). *Salters' Advanced Chemistry. Chemical Storylines*, Heinemann, Oxford, pp. 97–99.

21. Anonymous (1993). The ICI, Tioxide, Cleveland LEA, Teesside TEC Collaborative Learning Project. Education Liaison in Action, Issue 4, September 1993.

22. Anonymous (1992). The ICI, Tioxide, Cleveland LEA, Teesside TEC Collaborative Learning Project. Education Liaison in Action in the North East Region, Issue 2, September 1992.

23. Edgecombe, F.H.C. (1998). Regulations and practices of recycling in Nato Countries. B Canada and United States of America. Chapter 1: Introduction, In *Frontiers in the Science and Technology of Polymer Recycling*, ed. Akovali, G., Bernardo, C.A., Leidner, J., Utracki, L.A., and Xanthos, M., Kluwer, Dordrecht, pp. 29–39.

24. Sasse, F. and Emig, G. (1998). Chemical recycling of polymer materials, *Chem. Eng. Technol.*, **21**(10), 777–789.

25. Michaeli, W. and Breyer, K. (1998). Polymer recycling – status and perspectives. *Macromol. Symp.*, **135**, 83–96.

26. Rader, C.P. and Stockel, R.F. (1995). Polymer recycling: an overview. In *Plastics, Rubber and Paper Recycling*, American Chemical Society, Washington, DC, Chapter 1, pp. 2–10.

27. Eggles, P.G., Ansems, A.M.M. and van der Ven (2001). Eco-efficiency of recovery scenarios of plastic packaging. TNO Report (R2000/119) for AMPE, Apeldorn, The Netherlands.

28. Milgrom, J. (1994). Trends in plastics recycling, *Resour. Recycling*, **13**(5), 65–72.

29. Dennison, M.T. (1993). Plastics recycling: products, feedstock or energy? – a future view. Maack Conference Recycle 93, Davos, Switzerland.

30. Ellis, J.R. (1995). Polymer recycling: economic realities. In *Plastics, Rubber and Paper Recycling*, Chapter 5, *ACS Symp. Ser.*, **609**, 62–69.

31. Sutherland, G. and Tormey, M. (1996). A five-year history of recycling market prices: 1996 update, *Resour. Recycling*, **15**(10), 50–56.

32. Brandrup, J. (1998). Ecological and economical aspects of polymer recycling, *Macromol. Symp.*, **135**, 223–235.
33. Nunan, F. (1999). Barriers to the use of voluntary agreements: a case study of the Development of Packaging Waste Regulations in the UK, *Eur. Environ.*, **9**, 238–248.
34. DOE (1996). Producer Responsibility for Packaging – The Way Forward. New Release, No. 18 5, 7 May.
35. Truss, D.W. and O'Donnell, J.H. (1998). Plastics recycling: an Australian overview, *Macromol. Symp.*, **135**, 345–358.

5.7 REVISION EXERCISES

1. Summarise the main drivers for and barriers to recycling of polymer materials.

2. Examine Table 5.3 and discuss how different factors influence recycling of plastics. Compare how other factors influence the recycling of other materials and discuss the differences between recycling polymers and other materials.

Table 5.3

Recycling difficulties	Glass	Plastics	Paper/board	Metals	Composites
Capacity	X				
Output market/market price	X	X			
Contamination	X	X	X	X	
Supply-demand imbalance	X		X		
Insufficient amount of waste		X		X	X
Recycling lifetime					
Nature of waste (too thin)		X	X	X	
Recycling costs	X				
Noise	X				
Public participation	X	X	X	X	X
Nature of waste	X	X			
Recycling costs	X	X			

3. Why are the costs of collection and logistics higher than the costs of material reprocessing?

4. How does the price of oil influence the costs of polymer recycling?

5. When is the quality of the recyclate and issue? Why?

6. Compare the costs of PET recycling in the USA in 1993 with the current costs of PET recycling in Europe. What do you conclude? Has the situation changed much?

7. Do you participate in recycling? Why?

8. If you were an education officer, working for a Local Authority, what would you do to encourage the public to recycle? What would you do in that respect if you were a member of an environmental non-governmental organisation (NGO)?

9. How does socio-economic status and social conditioning influence public participation in recycling?

10. What do you think the plastics industry wants: increased recycling of material or energy recovery? Why?

11. Explain why you think incineration is a socially acceptable option in Denmark and Luxembourg, but not in the UK or Australia. Compare that with Japan.

12. Which policy measure would in your opinion be most effective in increasing polymer recycling rates?

13. Consider again question no. 7 in Chapter 4 and discuss how the choice of your integrated waste management strategy would change if, in addition to technical factors, you were also to consider logistics, costs, public acceptability and job creation opportunities. Answer this question by first considering developed countries and then by examining the options for the developed countries. Are the options different? Why?

Design for the environment: the life cycle approach

Chapter 6 – The Magic Circle (JW Waterhouse, 1875 – 83)
Waterhouse illustrates a witch casting a magic circle using a wand or anthame (a black-handled ceremonial dagger), while reciting a charm or spell, to purify and create a perimeter of space wherein evil magic is excluded. The circle, which was a mark of infinity and eternity, was represented by the "four elements": earth, air, water and fire – a fitting illustration of life cycle thinking and environmental considerations.

6.1 INTRODUCTION

One of the barriers to recycling polymers lies in the design of polymeric artefacts. Many products are very difficult to disassemble at the end of their useful life and that limits their recyclability or prevents their recycling altogether. Further difficulties are introduced with complex products, in which one or more types of polymer are commingled with other materials. In many cases it is almost impossible to separate polymers from the other materials and to distinguish between different types of polymer contained in these complex products. Typical examples include plastic materials contained in electrical and electronic goods or even in less complex products such as electrical cables.

Key Facts
● Designing
products 'from
cradle to grave'
could overcome
current difficulties
of identifying,
separating and
recovering the
components of
end-of-life
products.
● Life cycle
assessment seeks to
identify the
environmental
impacts of a
product from the
extraction of raw
materials, through
use/re-use to
eventual disposal.

One of the ways to contribute towards sustainability is to increase the recyclability of polymeric products and materials. However, for this to happen, the focus must shift from end-of-life waste management (discussed in this book so far) towards the front end of the product's life: its design. Adopting design-for-the-environment approaches would enable longer life cycles of polymeric products, either in the same application or in a cascade of different uses. This would also place the responsibility to reduce the environmental impacts of polymers on the producers rather than the users of the products. This is also consistent with the EU policy on producer responsibility (see Chapter 1), which puts an obligation on manufacturers to follow their products from 'cradle to grave'.

As we have seen in Chapter 1, the 'cradle to grave' concept is indeed central to the sustainability of products and materials. It enables environmental interventions to be tracked along the whole supply chains, from extraction of raw materials and their refining, through manufacture and use of a product to post-consumer waste management. In this way, a full picture of human interactions with the environment can be 'painted' to identify places where environmental improvements can be made. This concept is also known as 'life cycle thinking' because it follows a product and the associated environmental impacts throughout its life cycle. This broad system boundary presents a fundamental difference between life cycle thinking and other environmental analysis approaches, such as 'Waste Minimisation' and 'Waste Management', which consider only one or two stages in the life cycle. For instance, manufacturers often advertise the fact that their products and packaging are recyclable. However, that represents only one stage (waste management) in the life cycle of a product while other stages, such as the use or manufacture, could in the meantime be destroying the planet. Typical examples are deodorants and other consumer aerosols. Most of the aerosol cans carried a sign 'Recyclable' at the time when they still contained CFCs as propellants. Another example is the end-of-pipe abatement technologies: while they reduce emissions from a particular plant or manufacturing site, there may be a net increase in emissions and wastes arising elsewhere in the life cycle, due to the use of energy and materials for the abatement process.

It is thus obvious that broader, life cycle thinking is fundamental to the identification of more sustainable solutions. This is particularly true for polymeric products and materials because of their potential impacts on the environment from 'cradle' (non-renewable resource use) to 'grave' (solid waste management), including the intermediate stages such as production and use and the issue of additives, particularly endocrine (hormone) disrupters and dioxin-forming precursors (see Chapters 2 and 4). The following sections discuss the benefits of taking a life cycle approach, the way that it can be used for design for the environment and the tools that are necessary for this. The discussion is supported by the examples of polymeric packaging.

6.2 LIFE CYCLE THINKING: THE APPROACH AND THE TOOLS

The need to move away from narrow system boundaries and a limited view of the environment has led to the development of techniques and tools which incorporate life cycle thinking. One such tool is Life Cycle Assessment (LCA) which enables quantification and assessment of environmental performance of a product, process or an activity from 'cradle to grave'. Unlike other environmental management tools, such as Environmental Impact Assessment or Environmental Audit, which focus solely on the emissions and wastes generated by the plant or manufacturing site, LCA broadens system boundaries to consider environmental burdens and impacts along the whole life cycle of a product or a process. This holistic approach to environmental system management avoids shifting environmental burdens from one part of the system to another, as can often happen in a more narrow system analysis. For instance, prior to taking a life cycle approach, paper recycling had always been considered an activity that provided an overall benefit to the environment. The usual argument used was that it reduced solid waste and saved trees. However, LCA studies have shown that this is not always the case and that, for example, in Northern Europe paper incineration with energy recovery is environmentally a better option[1]. These at first surprising findings showed something that should have been obvious long before: recycling does not come without cost; it requires energy, materials, chemicals and transportation, all of which generate additional environmental impacts. Furthermore, the argument that paper recycling is 'good for the environment' because it saves trees does not apply in Northern Europe because trees are planted purposefully and forests are managed in a sustainable way. Results like these, obtained by adopting life cycle thinking and using LCA as a tool, have opened up a completely new debate on how resources should be used, re-used and recycled to the benefit of the whole society and the environment.

As illustrated in Figure 6.1, the life cycle of a product starts from extraction and processing of raw materials ('cradle'), which are then transported to the manufacturing site to produce a product. The product is then transported to the user and at the end of its useful life is either recycled and returned for reprocessing, or is disposed in a landfill ('grave'). The question mark in Figure 6.1 indicates that it may be possible to further extend the 'cradle to grave' concept and turn it into 'cradle to cradle'. This would require going back to our landfill sites and 'mining' valuable resources, which were discarded in the past as waste. Although at present this may sound like unfounded optimism, it is worth noting that this is already happening in the metals sector, particularly in North America, where some of the landfills are richer in metals content then some of the primary repositories[2]. Thus, it may be possible that in the future we will be able to close the materials loops completely and so reduce the use of primary resources. However, closing the loop in this way is not helped by the fact that we have mixed the waste prior to depositing it in the ground and that advanced separation technologies are needed if we want to reclaim them. Nevertheless, this may be a good reason to change our current waste management practices and start 'storing' different type of waste materials in separate landfill sites. Whatever cannot be re-used and recycled at present, may well become a valuable resource in the future, when more advanced technologies become available to reclaim them.

Key Facts
● A holistic approach to environmental system management avoids the shifting of environmental burdens from one part of the system to another.
● Recycling involves use of materials and energy and therefore itself has an environmental impact.
● Some landfill sites in the USA are being mined for valuable (metal) resources discarded in the past: the 'cradle-to-cradle' approach.

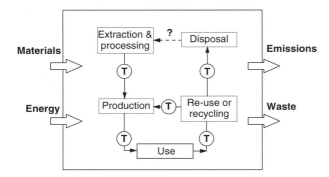

Figure 6.1 Stages in the life cycle of a product (T, transport)

The following section shows how the life cycle approach to environmental system management has been formalised into the LCA methodology.

6.2.1 LCA Methodology

LCA quantifies the use of materials and energy taken from the environment to generate goods and services within economic systems. It also identifies emissions and wastes that are associated with the life cycles of these goods and are eventually returned to the environment. This means that, in terms of the model of sustainable development discussed in Chapter 1 and illustrated in Figure 1.1, LCA as an environmental management tool can be positioned within the overlap between the economic and environmental components of sustainable development. This is shown in Figure 6.2. It also means that, if used correctly, LCA can help to identify more sustainable economic activities.

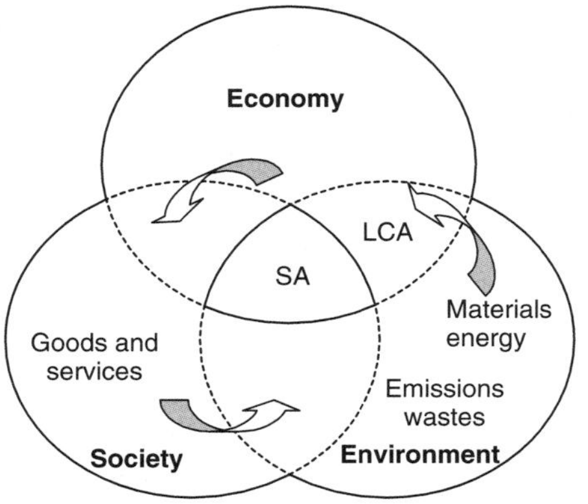

A correct use of LCA is indeed one of the most important issues associated with this type of analysis. Because LCA assumes a very broad view of usually complex systems (products, processes or activities) and considers a wide range of environmental impacts, it is quite possible for different LCA practitioners (or interest groups) assessing the same system to arrive at different conclusions. Some of the reasons for misinterpretation (and sometimes misuse) of LCA studies lie in the way the system boundaries are defined and in the type and quality of data used for analysis. Many LCA studies have been criticised and discredited for this reason, particularly if the results were used to gain commercial or other advantages.

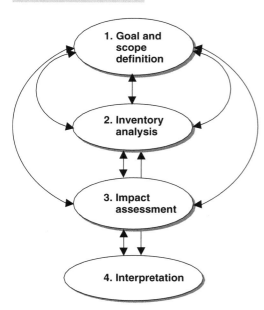

Figure 6.2 Positioning LCA as a tool for sustainable development (SA, sustainable activity)

Key Facts
● Different LCA studies can give 'different results' if the boundaries are defined differently and the data used are not consistent.

It is expected that standardisation of the LCA methodology, which has been finalised only recently, will help towards more uniform use of LCA and will contribute towards increasing its credibility as a tool. Two major international bodies have been involved in developing the methodology: Society for Environmental Toxicology and Chemistry (SETAC) and International Organisation for Standardisation (ISO). ISO 14040 series[3-6] defines four phases of LCA:

(1) Goal and Scope Definition (ISO 14041),
(2) Inventory Analysis (ISO 14041),
(3) Impact Assessment (ISO 14042),
(4) Interpretation (ISO 14043).

Figure 6.3 shows the position and interactions of these phases within the LCA methodological framework.

Goal and Scope Definition

The first and probably most critical phase of an LCA study is the goal and scope definition. This component includes defining the purpose of the study and its intended use, *i.e.* whether the study is going to be used internally by a company for improving the performance of their system or externally, *e.g.* for marketing or influencing public policy. Defining the system and system boundaries determines the scope of the study. The assumptions and limitations of the study are also identified in this phase.

It must be borne in mind that the LCA system boundary should be drawn to encompass all stages in the life cycle from extraction of raw materials to the final disposal. As already explained, this is referred to as a 'cradle-to-grave' approach. However, in some cases, the scope of the study will demand a different approach, where it is not appropriate to include all stages in the life cycle. For instance, this is the case with intermediate products (*e.g.* granulated polymer), which can have a number of different uses, so that it is not possible to follow their

Figure 6.3 The methodological framework for LCA

1. Goal and scope definition

2. Inventory analysis

3. Impact assessment

4. Interpretation

numerous life cycles after the manufacturing stage. The scope of such studies is from 'cradle to gate', and they follow a product from the extraction of raw materials to the point where they leave the factory gate.

The functional unit, one of the most important elements of an LCA study, is also defined in this phase. This is a quantitative measure of the output of products or services that the system delivers. In comparative studies it is crucial that systems are compared on the basis of equivalent function, *i.e.* functional unit. For example, comparison of different drinks packaging should be based on their equivalent function, which is to contain a certain amount of beverage. The functional unit is then defined as 'the quantity of packaging necessary to contain a specified volume of beverage'. This will be further explained in the context of a packaging polymer example later in Section 6.2.2.

This phase also includes an assessment of the data quality. As indicated in Figure 6.3, the goal and scope are constantly reviewed and refined while LCA is being carried out, as additional information and data become available.

Inventory Analysis

Life Cycle Inventory (LCI) analysis represents a quantitative description of the system and is thus the most objective phase in LCA. Inventory Analysis includes:

- further definition of the system and its boundaries;
- flow diagrams of the systems;
- data collection;
- allocation of environmental burdens;
- calculation and reporting of the results.

Following a general definition in the Goal and Scope Definition phase, the system is further defined and characterised in LCI to identify the data needs. A system is defined as a collection of materially and energetically connected operations (including for example manufacturing process, transport or fuel extraction process) which performs some defined function. The system is separated from its surroundings, *i.e.* the environment, by a system boundary. Thus, for these purposes the environment is defined along with the system, by exclusion. This simple definition is illustrated by Figure 6.4.

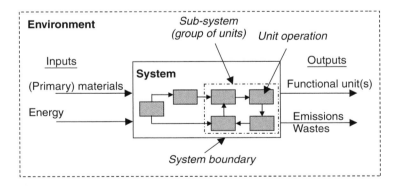

Figure 6.4 Inventory Analysis: a flow diagram showing a system and its boundaries

The system is disaggregated into a number of inter-linked subsystems and their interdependence is illustrated by flow diagrams (see Figure 6.4). Depending on the level of detail of the available data, the subsystems can represent the unit operations or a group of units. Each subsystem is described in detail by flows of materials and energy, as well as emissions to air and water and solid wastes. All inputs into, and outputs from, the

Key Facts
- 'Cradle-to-gate' analyses assess the impacts of a product from extraction to the point where it leaves the factory gate for a possible myriad of uses.
- The functional unit describes the function of a system under the study and enables comparisons of different systems on the basis of the equivalent function that they deliver.
- A system is defined as a collection of materially and energetically connected operations which perform a defined function.
- Mass and energy balances are carried out for the system under study in the LCI.

subsystems are balanced in this phase and data are normalised with respect to the unit output from each subsystem. This is equivalent to carrying out mass and energy balances, an approach central to process systems analysis. On the basis of the data collected for a period statistically relevant for the study (*e.g.* 1 year), the environmental burdens defined as resource depletion and emissions to the environment, are then calculated for the whole system. The results are listed in the inventory tables and represented graphically. Environmental burdens include, for instance, fossil fuel consumption, emissions of sulphur dioxide, emissions of metals to water and amount of solid waste, and they can be calculated as:

$$B_k = \sum_{i=1}^{I} bc_{k,i} x_i \qquad k = 1, 2, \ldots, K \tag{6.1}$$

where $bc_{k,i}$ is burden k associated with the material or energy flow x_i in a process or activity. An example would be an emission of CO_2 (burden $bc_{k,i}$) generated *per* tonne of natural gas (material flow x_i) used to generate electricity (process or activity). As defined by Equation (6.1), there would be in total K burdens from I flows. A simple example in Textbox 6.1 illustrates how the burdens can be calculated.

The Inventory Analysis phase also deals with allocation of environmental burdens, the problem encountered in multiple-function systems, such as co-product systems, waste treatment and recycling. Allocation is the process of assigning to each function of a multiple-function system only those environmental burdens that each function generates. An example of a co-product system is a naphtha cracker that produces ethylene, propylene, butenes and pyrolysis gasoline, all of which can be used as raw materials to produce various types of polymer (see Chapter 3). The allocation problem here is to assign to each of the products or functional outputs only those environmental burdens for which each product is 'responsible'. The usual approach is to use either mass or economic basis, allocating the total burden according to the mass output or economic value of each product or functional output. This is illustrated in Textbox 6.2.

The simple example in Textbox 6.2 demonstrates that the allocation method will usually influence the results of the study so that the identification of an appropriate allocation method is crucial. To guide the choice of the correct allocation method, ISO 14041[4] recommends the following three-step procedure:

(i) If possible, allocation should be avoided by expanding the system boundaries or disaggregating the given process into different subprocesses.
(ii) If it is not possible to avoid allocation, then the allocation problem must be solved by using system modelling based on physical causation which reflects the underlying physical relationships among the functional units.
(iii) Where physical relationships cannot be established, other relationships, including economic value of the functional outputs, can be used.

Further reading on allocation and application to real industrial examples can be found in a series of papers by Azapagic and Clift (see further reading at the end of the chapter).

Following data collection, allocation and calculation of total burdens in the system, the results are presented in inventory tables and accompanied by graphs. Comparison of alternative products, processes or activities for which LCA is being performed may be carried out at this stage and some conclusions drawn on which is the preferred alternative. However, a large number of burden categories makes it often difficult to make any definite decisions at this stage and hence it may be helpful to proceed to the Impact Assessment stage.

Key Facts
- Environmental burdens are calculated in terms of the consumption of renewable and non-renewable resources, emissions to air and water and generation of solid waste.
- In multiple-function systems, the environmental burdens must be allocated between the different functions.
- Allocation can be done on a mass basis or on the basis of economic value of the functional outputs.

To illustrate how the environmental burdens and impacts from a system can be calculated, let us consider a hypothetical system whose function is to contain and deliver 1000 l of water in 1-litre polypropylene (PP) bottles. The functional unit is then defined as the amount of PP needed to make 1000 bottles. In our case, each bottle weighs 50 g so that the functional unit is defined as:

50 g *per* bottle $\times$ 1000 bottles $=$ 50 kg

For the purposes of illustration, the system can be divided into the following subsystems: extraction and refining of crude oil, production of PP bottles, use of bottles (or, rather, water) and disposal of empty bottles in a landfill. This is illustrated by a simple flow diagram below. Let us assume that each flow i will be associated with a certain amount of CO_2 and CH_4 emissions. For example, 'Extraction, Refining and Cracking' generate a flow of $x_1 = 52$ kg of propylene *per* functional unit (F.U.), *i.e. per* 1000 bottles. This is associated with the emissions of 0.1 kg of CO_2 and 4×10^{-4} kg of CH_4 *per* kg of x_1. The CO_2 and CH_4 emissions from 'Bottle Delivery and Use' are generated from bottle transport to the point of use (retailer). The functional unit is represented by $x_3 = 50$ kg, the output from 'Bottle Delivery and Use'. Note that the flows are not mass balanced; also note that 10 % of bottles are not collected for disposal, hence only 45 kg PP reaches the landfill.

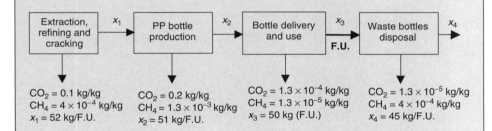

Using Equation (6.1), the total environmental burdens *per* functional unit related to the emissions of CO_2 and CH_4 are therefore equal to:

$$B_{CO_2} = \Sigma \, bc_{CO_2} x_i = 0.1 \times 52 + 0.2 \times 51 + 1.3 \times 10^{-4} \times 50$$
$$+ 1.3 \times 10^{-5} \times 45 \Rightarrow B_{CO2} = 15.4 \text{ kg } per \text{ F.U.}$$
$$B_{CH_4} = \Sigma \, bc_{CH_4} x_i = 4 \times 10^{-4} \times 52 + 1.3 \times 10^{-3} \times 51$$
$$+ 1.3 \times 10^{-5} \times 50 + 4 \times 10^{-4} \times 45 \Rightarrow B_{CH_4} = 0.1 \text{kg } per \text{ F.U.}$$

The global warming potential (GWP) associated with these two greenhouse gases can be calculated by applying Equation (6.2) and the classification factors for CO_2 (1 kg kg^{-1}) and CH_4 (11 kg kg^{-1}) given in the table in Appendix 2:

$$E_{GWP} = ec_{CO_2} B_{CO_2} + ec_{CH_4} B_{CH_4} = 1 \times 15.4 + 11 \times 0.1$$
$$\Rightarrow E_{GWP} = 16.5 \text{ kg } per \text{ F.U.}$$

Textbox 6.2 Allocation of environmental burdens in Inventory Analysis: an example

Consider a co-product system producing two products or functional outputs, Product 1 and Product 2, and generating an emission of CO_2. A simple flow diagram of this system is shown below. The allocation problem is related to identifying a correct method for apportioning the total CO_2 emissions to each product. For the purposes of illustration only, the two most often used approaches for allocation are shown here:

(i) allocation based on the products mass outputs, x_1 and x_2;
(ii) allocation based on their economic value, f_1 and f_2.

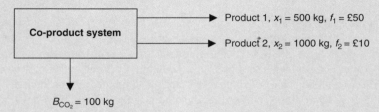

Following the notation in Equation (6.1), the allocation coefficients are defined by $bc_{k,i}$ so that we have:

(i) Allocation on mass basis:

$$\text{Product } 1: bc_1 = [x_1/(x_1 + x_2)]B_{CO_2} = [500/(500 + 1000)] \times 100$$
$$\Rightarrow bc_1 = 33.3 \text{ kg } CO_2$$
$$\text{Product } 2: bc_2 = [x_2/(x_1 + x_2)]B_{CO_2} = [1000/(500 + 1000)] \times 100$$
$$\Rightarrow bc_2 = 66.7 \text{ kg } CO_2$$

(ii) Allocation on the basis of economic value:

$$\text{Product } 1: bc_1 = [f_1/(f_1 + f_2)]B_{CO_2} = [50/(50 + 10)] \times 100$$
$$\Rightarrow bc_1 = 83.3 \text{ kg } CO_2$$
$$\text{Product } 2: bc_2 = [f_2/(f_1 + f_2)]B_{CO_2} = [10/(50 + 10)] \times 100$$
$$\Rightarrow bc_2 = 16.7 \text{ kg } CO_2$$

Obviously, the allocation coefficients obtained by these two different methods are quite different for the same functional outputs. For example, the mass-based approach allocates 33.3 % of the total CO_2 emissions to Product 1, while allocation based on the economic value of the products assigns 83.3 % of the total CO_2 burden to the same product. This means that in this case the use of the two different allocation methods would lead to completely different results from the LCA study. This simple example illustrates the importance of using the appropriate allocation approach, depending on the type of system analysed. A guide to allocation for different types of system can be found in ISO 14041[4].

Impact Assessment

The effects of environmental burdens identified in Inventory Analysis are assessed in Impact Assessment. This part of LCA is based on both quantitative and qualitative procedures to characterise and assess the environmental impacts of a system. The burdens are first aggregated into a smaller number of impact categories to indicate the potential impacts on human and ecological health and on resource depletion. The aggregation is undertaken on the basis of potential impacts of the burdens, so that one burden can be associated with a number of impacts; *e.g.* VOCs contribute to both global warming and ozone depletion. The approach used most widely for classification of impacts is known as 'problem oriented', whereby the burdens are aggregated according to their relative contributions to the environmental effects that they may have. The impacts most commonly considered in LCA are:

- resource depletion,
- global warming,
- ozone depletion,
- acidification,
- eutrophication,
- photochemical (summer) smog,
- human toxicity,
- aquatic toxicity.

The definitions of these impacts are given in Appendix 2. Further details on the problem-oriented approach and impacts can be found in Heijungs *et al.*[7]

In the problem-oriented approach, the impacts are calculated relative to a reference substance. For instance, CO_2 is a reference gas for determining the global warming potential of other related gases, such as CH_4 and VOCs. In general terms, impact E_l can be calculated by using the following formula:

$$E_l = \sum_{k=1}^{K} ec_{l,k}B_k \qquad l = 1, 2, \ldots, L \qquad (6.2)$$

where $ec_{l,k}$ represents the relative contribution of burden B_k to impact E_l. The calculation procedure for different impact categories is given in Appendix 2. A simple illustration of the calculation of global warming can be found in Textbox 6.1.

The impacts are sometimes normalised on the total impacts in a certain area over a given period of time. Some argue that, since LCA is global in its character, total world annual impacts should be used as the basis for normalisation. Total emissions of global warming gases and world resource depletion can be calculated relatively easily. However, other impacts, such as acidification or human toxicity, are more difficult to determine on the global level so that normalisation is still not a reliable method for evaluating total impacts of a system.

Although the number of impact categories is much smaller than the number of burdens identified in the inventory analysis, it is still significant, which can make comparison of alternative systems difficult, particularly when one system is better in some but worse in the other impacts. In an attempt to aid decision-making, some people advocate further aggregation of impacts into a single environmental impact function by assigning weights of importance to different impacts. This process, known as Valuation, reduces eight or so impact categories into a single number, *EI*, as represented by the following formula:

$$EI = \sum_{l=1}^{L} w_l E_l \qquad (6.3)$$

where w_l represents relative importance of impact E_l. For example, on a scale of 1 to 10, each impact can be assigned a score (or weight) w_l from 1 to 10 to indicate its importance in relation to other impacts; the higher the score the higher the 'importance' of the impact to the decision-makers.

A number of techniques have been suggested for use in Valuation. They are mainly based on expressing preferences either by decision-makers, 'experts' or the public. Some of these methods include multiattribute utility theory, analytic hierarchy process, impact analysis matrix, cost–benefit analysis and contingent valuation. However, because of a number of difficulties on both philosophical and practical levels associated with using these techniques, there is no consensus at present on how to aggregate the environmental

Key Facts
- Environmental burdens are aggregated into a smaller number of environmental impacts according to their relative contributions to the environmental effects that they may have relative to a reference substance (such as CO_2 for determining global warming potential).
- Valuation reduces all environmental impacts to a single value, by assigning weights of importance to each, but the result is subjective and may lead to different interpretations of LCA study.

impacts into a single environmental impact function. Among others, Fava *et al.* discuss the valuation methods in more detail (see list for further reading).

Interpretation

The Interpretation phase is aimed at system improvements and innovation. In addition it also covers the following steps: identification of major burdens and impacts, identification of stages in the life cycle that contribute the most to these impacts (so-called 'hot spots'), evaluation of these findings, sensitivity analysis for data quality and gaps and final recommendations.

It is important to note that data quality is fundamental to LCA. Some of the data quality issues such as reliability and consistency can be overcome by using standardised databases, which are starting to emerge after years of data compilation and their incorporation into publicly and commercially available databases. A range of LCA software packages is now also available and they comprise more or less reliable databases on materials, energy, transport and waste management options. Examples include TEAM and DEAM[8], PEMS[9] and SimaPro[10] LCA software.

Further detail on LCA methodology can be found in Consoli *et al.* (see further reading list) and in the ISO 14040 series[3-6]. The example of plastic packaging illustrated in the next section shows how LCA can be performed following the methodology discussed above.

Key Facts
● Identifying the 'hot spots' in the system through LCA can help find optimum options for system improvements.

6.2.2 LCA of Polymeric Packaging: An Example

Goal and Scope Definition

The goal of this LCA study is to compare the life cycle environmental impacts of water bottles manufactured from four different polymeric materials: high density polyethylene (HDPE), low density polyethylene (LDPE), polypropylene (PP) and high impact polystyrene (PS). The scope of the study is from 'cradle to grave', following the bottles from the extraction and processing of raw materials through the manufacture to their final disposal, including transportation. For the purposes of illustration, the functional unit is defined as 'the amount of packaging needed to contain 1000 l of water in bottles of 1 l'.

Inventory Analysis

In addition to the polymers used to make the bottles, two more materials are used in the system: PP for caps and paper for labels. It is assumed that each polymer is transported to the manufacturing site to form the bottles, which are then filled. At this stage, the PP caps and paper labels (PL) are also added. The bottles are then transported to the use phase. After use, the majority of waste bottles are disposed of in a landfill, while a small number are incinerated with energy recovery (electricity). The system is credited by subtracting this amount of electricity recovered from the total electricity requirement in the system. The emissions generated during incineration are also accounted for. Figure 6.5 shows the LCA flow diagram of the bottle system with different polymers used in their manufacture. Figure 6.6 follows the production of the polymers from 'cradle to gate', *i.e.* from extraction of fossil fuels to the point where they leave the factory gate, to be transported to the bottle manufacturing site. PP caps and PL are also tracked from 'cradle to gate', before they enter the system shown in Figure 6.5.

It should be noted that there are four systems shown in Figure 6.5, each corresponding to a different type of polymer used to manufacture the bottles. Since for the purposes of this analysis the systems are identical in all elements except for the polymeric materials, the comparison is in effect between different types of polymers. Generally, if the purpose of the study is to compare different systems that have some identical elements, then these

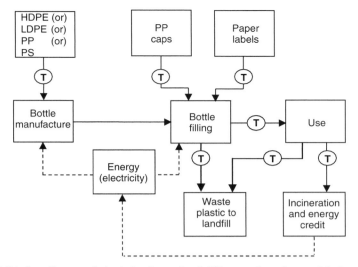

Figure 6.5 LCA flow diagram of plastic bottles made of different polymeric materials (⎯⎯ material flows; - - - energy flows, T – transport)

Key Facts ⚷⎯⎯
● In comparing systems with identical elements, these elements can be disregarded in the analysis.

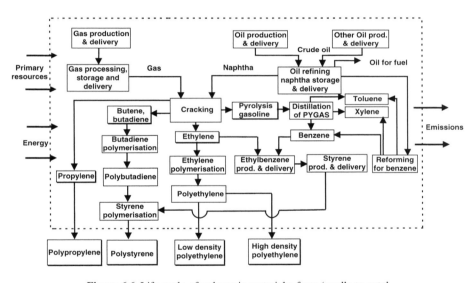

Figure 6.6 Life cycle of polymeric materials: from 'cradle to gate'

can be disregarded in the analysis. However, for completeness and for illustration of the LCA methodology, the full systems are analysed here.

Based on the definition of the functional unit, the amounts of materials entering the use phase are as follows:

HDPE: $x_{HDPE} = 47\,g$ *per* bottle $\times$ 1000 bottles $\Rightarrow x_{HDPE} = 47\,kg$

LDPE : $x_{LDPE} = 45\,g$ *per* bottle $\times$ 1000 bottles $\Rightarrow x_{LDPE} = 45\,kg$

PP : $x_{PP} = 44\,g$ *per* bottle $\times$ 1000 bottles $\Rightarrow x_{PP} = 44\,kg$

PS : $x_{PS} = 51\,g$ *per* bottle $\times$ 1000 bottles $\Rightarrow x_{PS} = 51\,kg$

Each type of bottle requires the same amount of PP caps and paper labels:

PP caps : $x_{PP-C} = 2\,g$ *per* bottle $\times$ 1000 bottles $\Rightarrow x_{PP-C} = 2\,kg$

PL : $x_{PL} = 1\,g$ *per* bottle $\times$ 1000 bottles $\Rightarrow x_{PP-C} = 1\,kg$

It has been assumed that the electricity requirement for bottle forming is $3.6\,MJ\,kg^{-1}$ of polymer; to fill 1000 bottles, 53 MJ of electricity is used. The polymers are transported a distance of 500 km by road and rail to the manufacturing site and the full bottles travel 300 km by road to the retailer (use phase). Caps and labels each travel 100 km by road to the bottle filling site; waste bottles are transported 25 km by road to either landfill or incineration.

The life cycle inventory data used in this example have been obtained from publicly available databases. The inventory results comparing the four types of plastic bottles are shown in Figure 6.7. For illustration purposes only four burdens are shown: energy use, oil reserves, and CO_2 and NO_x emissions. In real case studies, a large number of burdens, comprising over 100 different categories, would usually be considered.

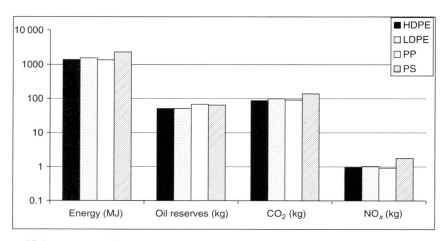

Figure 6.7 Inventory analysis: comparison of environmental burdens for water bottles made from different polymers

The results in Figure 6.7 show that the PP bottles are preferred over other bottles with respect to two burden categories: energy use and NO_x emissions, while LDPE bottles are marginally favoured for oil reserves and HDPE for CO_2 emissions. However, the difference between PP, HDPE and LDPE bottles, in terms of energy use and CO_2 and NO_x emissions, is small (around 3%) so that the only burden that sets them apart is that of oil reserves, where LDPE and HDPE use around 25% less oil than PP (note the use of a logarithmic y axis). For this burden, PP is the worst option ahead of PS, which is the least preferred option for the remaining three burdens.

The same analysis would be done for the remaining burdens not shown here. However, as already suggested, choosing the best option based on a number of burden categories may not be an easy task, particularly if the preferred option changes from burden to burden. The aggregation of a large number of burdens into a smaller number of impacts in Impact Assessment reduces significantly the number of categories that have to be considered, which may facilitate the decision-making process. The Impact Assessment results are discussed below.

Impact Assessment

Figure 6.8 compares the bottles for eight impact categories (see Appendix 2 for definitions). These results, expressed *per* functional unit, show that PS is the worst of the four polymers for all impact categories. For many categories, the difference between PS and the best option is over 50%, with aquatic toxicity being the highest at 78%. Thus, the choice will have to be made among the three remaining polymers. This is part of the Interpretation phase.

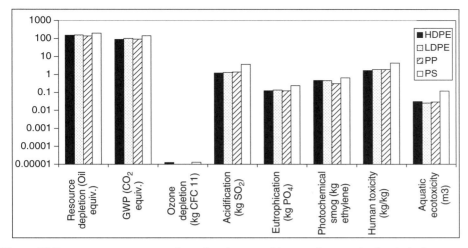

Key Facts 🔑
● Setting a value to
an environmental
impact is a
subjective choice,
strongly influenced
by one's personal
viewpoint.

Figure 6.8 Impact assessment: comparison of environmental impacts for water bottles made from different polymers

Interpretation

Given that each of the three polymers is better for some impacts but worse for the others, choosing the best option is not easy. This is one of the typical situations in environmental decision making, and often encountered in LCA, where one option is preferred for some impacts and not for the others. As discussed in the section on LCA methodology, some people prefer to carry out Valuation, by assigning weights of importance to the impacts (see Equation 6.3). However, it is obvious that different people will have different views on the importance of the impacts, depending on many factors, including their location and personal concerns. For example, people living in California may attach the highest weight to photochemical (summer) smog, which in this case study would favour PP. On the other hand, the Scandinavians may consider acidification as the most important impact in which case HDPE would be their preferred option. It is thus apparent that deriving a universal set of weights of importance for different impacts would be neither possible nor meaningful.

One of the ways to tackle decision-making problems with multiple and often conflicting objectives is to allow the decision-makers to analyse the disaggregated results, as shown in Figure 6.8, and to 'trade off' the impacts. For example, LDPE is the preferred option in one of the impacts only: aquatic toxicity. One possible question is: how significant is the difference between this value and the next preferred option, in this case PP? The answer is 12%, which may be significant, so no decision can be made yet on whether to eliminate LDPE from further considerations.

The next step in this case would be to consider the remaining two polymers. The analysis reveals that PP is preferred for three categories: resource depletion, eutrophication and photochemical smog. HDPE scores better for the other three impacts: global warming, acidification and human toxicity. All three polymers are responsible for approximately the same level of ozone depletion. Analysing the relative difference in impacts between these two polymers shows that, for global warming and eutrophication, they only differ by 2% and 5% respectively, which could be argued to be not significant. That leaves four impacts for further consideration. PP uses 11% fewer resources and generates 38% less photochemical smog than HDPE. On the other hand, the values for acidification and human toxicity generated by HDPE are 15% lower compared to PP. It should be noted that for acidification, PP is the third preferred choice, after LDPE. Table 6.1 summarises the above considerations and ranks the options in order of preference from 1 to 4, with 1 corresponding to the best and 4 to the worst option for a particular impact. The difference between the best and second best option is also shown.

Table 6.1 Ranking the options.

Impact	HDPE	LDPE	PP	PS	Difference between 1st and 2nd option (%)
Resource depletion	2	3	1	4	11
Global warming	2	3	1	4	2
Ozone depletion	3	2	1	4	1
Acidification	1	2	3	4	15
Eutrophication	2	3	1	4	5
Photochemical smog	3	2	1	4	35
Human toxicity	1	2	3	4	15
Aquatic toxicity	3	1	2	4	12

Key Facts 🐭━━━━🔑
● The alternative to valuation is to rank different options for each environmental impact in the order 'best to worst' and to evaluate the difference between the best and the other options.
● In this way decision makers can make a decision based on a consistent and transparent set of data.

The decision-makers now have a clearer picture as to the ranking of the options, and are able to 'trade off' impacts and thus make a decision that will represent their interests and concerns in the best way. Clearly, different decision-makers and interest groups will make different decisions based on this information, so no attempt is made here to suggest what the best option might be. The important point to note, however, is that decision-making structured in this way is transparent and easy to understand and justify. It also avoids the problems encountered in Valuation, which requires articulating preferences for impacts before the trade-offs have been explored and before decision-makers understand what they can gain or lose by choosing different options.

Interpretation of LCA results also involves evaluation of options for system improvements. Identification of 'hot spots' helps to identify stages in the life cycle with highest impacts so that these stages can be targeted for maximum improvements. As an illustration, Figure 6.9 shows the 'hot spots' in the life cycle of HDPE bottles for the burdens discussed in Inventory Analysis. The total values of the burdens are also shown. There are two 'hot spots' in this system: HDPE production (comprising polymer production from 'cradle to gate' and bottle manufacture) and electricity generation. These two stages contribute to the majority of CO_2 emissions in the bottle system. HDPE production and bottle manufacture are also responsible for depletion of most of the oil reserves used in the system and NO_x emissions. The other contributors to CO_2 emissions are incineration and transport. However, incineration is also a 'cool spot' for oil reserves and NO_x emissions: it reduces the overall impacts (shown as negative values) from the system by displacing fossil fuel-based electricity generation.

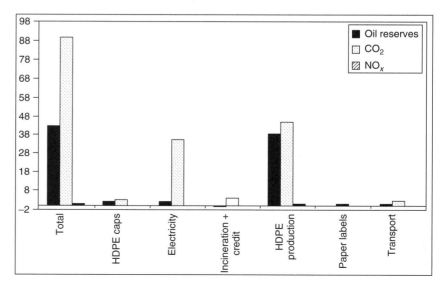

Figure 6.9 Identifying 'hot spots' for HDPE bottles

This simplified case study illustrates what kind of information can be obtained through an LCA study. Firstly, it can identify and quantify the major burdens and impacts along the life cycle of a product. Secondly, it can identify the 'hot spots' in the system, showing which life cycle stages contribute most to these impacts and should therefore be targeted for improvements. Finally, LCA provides information to consumers, manufacturers and other decision-makers to enable identification of environmentally more sustainable products, processes or activities.

6.3 DESIGN FOR THE ENVIRONMENT: LIFE CYCLE PRODUCT DESIGN

Historically, most of the LCA literature and case studies have been product-oriented. The literature body on product LCAs is vast and its review is outside the scope of this book. Some examples for further reading on product LCA case studies include:

- agricultural products: Audsley *et al.* (1997), Haas *et al.* (2000);
- chemicals: Ophus and Digerness (1996), Dobson (1996), Franke *et al.* (1995);
- electronic products: Miyamoto and Tekawa (1998), de Langhe *et al.* (1998);
- food and drinks: Cederberg (1998), Anderson and Ohlsson (1999), Bell *et al.* (2000);
- gases: Aresta and Tomassi (1997), Rice (1997);
- metals and minerals: Robertson *et al.* (1997), Chubbs and Steiner (1998), Azapagic and Clift (1999c);
- paper products: Seppala *et al.* (1998), Backlund (1998);
- polymers: Boustead (1992), APME (1992–1997), Yoda (1996);
- textiles and leather: Kuusinen *et al.* (1998), Puntener 1998, Beck *et al.* (2000).

The potential for using LCA as a tool for process evaluation has been recognised only relatively recently and this has led to the development of life cycle approaches for process selection and optimisation. An extensive review of process-oriented LCA applications can be found in the paper by Azapagic (1999) in the list of references for further reading.

One of the newly emerging applications of LCA is in product design, which has resulted in the development of an LCA-based tool: Life Cycle Product Design (LCPD). The LCPD methodology is still developing and several approaches have been proposed by different authors[11–15]. A general methodological framework for LCPD[12,13] is presented below, with an application to polymeric materials. It should be noted that a similar methodology can also be applied to process design and optimisation. Further reading on Life Cycle Process Design may be found in the list for further reading.

6.3.1 Methodological Framework for LCPD

The methodological framework for LCPD[12,13] is outlined in Figure 6.10. LCA is used throughout the product development procedure, initially on a reference product. This enables quantification of the main environmental impacts and stages in the life cycle that contribute most to these impacts. It also enables the identification and evaluation of options for environmental improvements of the product. The improvements are then achieved through selection of the best materials and technologies to achieve minimum environmental impacts. In making the choices, the whole life cycle of the materials and technologies is considered, ensuring that the burdens are not merely shifted from one part of the system to another. In this way, the best available environmental options are identified.

However, as the product has to satisfy a number of other criteria, the design cannot be based solely on environmental criteria. The first criterion for consideration in designing a new or improving an existing product is its technical performance: if the product does not satisfy the performance standards then it cannot deliver the function for which it is designed. Secondly, the product must be economically viable or its production will not be feasible. Furthermore, the product and its production must comply with relevant legislation, including health and safety regulations and environmental emission limits. The supply chain is a further criterion that is considered within this methodological framework that

Key Facts 🔑
- The first consideration in designing or improving a product is that it meets key performance criteria.
- Other important design criteria are economic viability, conformance to health and safety and environmental legislation, quality of component parts and customer preferences.

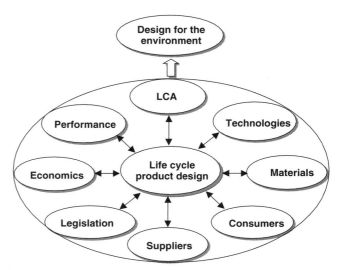

Key Facts
● The best life cycle
product design
(LCPD) is achieved
by an iterative
approach.
● All possible life
cycles and cascaded
uses are considered
with LCPD, using
multi-objective,
mathematical
optimisation
methods.

Figure 6.10 A general methodological framework for Life Cycle Product Design (LCPD). Reproduced with permission from Azpagic (1997). 'Life cycle assessment: a tool for innovation and improved environmental performance.' In '*Science, Technology and Innovation Policy.*' Conceicao *et al.* (eds). Part **VI.35**, pp. 519–530. Copyright (1997) Quorum Books, Westport, USA

enables identification of best suppliers, in terms of their product quality, environmental and other performance criteria. For instance, one of the guiding principles in choosing the best supplier could be their certification to the ISO 9000 (product quality) or 14000 (environmental management system) standards. Finally, consumers and their preferences must also be taken into account.

Once all of these requirements have been considered and met, LCA is performed again to identify and quantify the improvements made. This whole process is iterative with a continuous exchange of information and can yield a number of possibilities for improvements.

Therefore, LCPD offers a potential for technological innovation in the product concept and structure through selection of the best material and process alternatives over the whole cycle. This can be of particular importance if placed within the context of the ISO 14000 Environmental Management Systems (EMS) and the EU Integrated Pollution Prevention and Control (IPPC) Directives, which require companies to have a full knowledge of the environmental consequences of their actions, both on and off site. Furthermore, 'producer responsibility' and 'take back' initiatives are starting to exert pressure on manufacturers to reduce waste at source and manage the post-consumer waste associated with their products (as discussed in Chapter 1). LCPD can provide a powerful framework for the design of products that are easy to disassemble, reprocess and recycle. This, in turn, aids planning the re-use of the materials in the same life cycle as well as in the 'cascaded' use of resources, whereby the materials pass from one life cycle to another to be recycled in a number of different uses. For example, a plastic material can be re-used in one life cycle as a bottle, which at the end of its life cycle can be used to produce a plastic crate, which may then be processed to make fibres for carpet lining. Therefore, the waste from one system or life cycle becomes an input material into another. As we have seen in Chapter 1, this concept is often referred to as 'industrial ecology' of materials or 'industrial symbiosis'. Further reading on industrial ecology is provided by Graedel and Allenby.

However, given the range of criteria that need to be considered within the LCPD framework and a number of possible life cycles or cascades of uses, it soon becomes clear that identification of optimum options is not a trivial task. To aid this process, a robust mathematical modelling and optimisation framework is required. Owing to the multi-objective nature of the problem, in which optimum solutions are sought for a number of, often conflicting, criteria, it is appropriate to use multi-objective optimisation. This enables simultaneous optimisation on a number of objective functions, subject to the constraints in the system. For example, objectives to be optimised include environmental impacts (*e.g.* global warming, ozone depletion, acidification) and economic costs. Constraints are defined by performance criteria, availability of raw materials, legislation, *etc*. More detail on multi-objective optimisation in the context of LCA can be found in the further reading. Further discussion here shows how the approach to life cycle-based system modelling and optimisation[12-16] can be used for identification of optimum options for product design and cascaded use of materials. The approach is illustrated by an example of plastic packaging. Less mathematically inclined readers can skip this section and continue reading from Section 6.3.2.

Key Facts
● LCPD requires consideration of a number of objectives and criteria.
● Multiobjective optimisation can help identify a range of optimum solutions.
● Decision makers then have to trade-off different objectives to choose the 'best compromise' solutions.

Mathematical Modelling and Optimisation Framework for LCPD

System optimisation in the context of LCA amounts to minimising the environmental burdens or impacts from 'cradle to grave', subject to certain constraints. However, as discussed above, decisions are not made on the basis of environmental criteria only; a number of technical, financial and social factors must also be considered. Therefore, in the context of LCPD and cascaded use of materials, optimisation is performed on a number of functions, including environmental and socio-economic objectives. The optimum solutions are defined by a multidimensional noninferior or Pareto* surface. By definition, none of the objective functions at the Pareto optimum can be improved without worsening some other objective function. Therefore, some trade-offs between objective functions are necessary in order to reach the preferred optimum solution in a given situation. Thus, several alternative solutions are obtained; they are all optimal, but the choice of the best one will depend on a range of technical, financial, environmental and social criteria considered by the decision-makers. In this way, acceptable solutions, representing a compromise between conflicting objectives, can be found.

In an LCA-based optimisation model, the system is optimised on environmental objectives, defined as burdens and given by Equation (6.1) in the section on Inventory Analysis:

$$\text{Minimise} \quad B_k = \sum_{i=1}^{I} bc_{k,i} x_i \quad k = 1, 2, \ldots, K \tag{6.1}$$

where $bc_{k,i}$ is burden k associated with a material or energy flow x_i encompassing all activities from 'cradle to grave'. The objective functions may also be defined as environmental impacts given by Equation (6.2) in the section on Impact Assessment:

$$\text{Minimise} \quad E_l = \sum_{k=1}^{K} ec_{l,k} B_k \quad l = 1, 2, \ldots, L \tag{6.2}$$

where $ec_{l,k}$ represents the relative contribution of burden B_k to impact E_l; for example, global warming potential factors, $ec_{l,k}$, for different greenhouse gases are expressed relative to that of CO_2.

Within the LCPD framework, the system is also optimised on the economic objectives, such as profit or costs, as given by:

$$\text{Maximise (or minimise)} \quad F = \sum_{i=1}^{I} f_i x_i \tag{6.4}$$

Optimisation is performed simultaneously on objectives (6.1), (6.2) and (6.4), subject to constraints which are usually defined by material and energy balances, productive capacities, resource availability, market demand, legislative limits and so on:

$$\sum_{i=1}^{I} a_{j,i} x_i \leq A_j \quad j = 1, 2, \ldots, J \tag{6.5}$$

* Pareto was a new welfare economist who defined a social state as (Pareto) optimal if no individual can be made better off without making at least one other individual worse off. In other words, if such a state is reached it is not possible to increase the utility of some individuals or groups without diminishing that of others.

where $a_{j,i}$ is an input or output coefficient corresponding to flow x_i in the activities from 'cradle to grave'. Equations (6.4) and (6.5) can be defined as linear or nonlinear; Equations (6.1) and (6.2) are currently usually represented by linear relationships because of the way the burdens and impacts are defined in LCA.

The optimisation procedure within the LCPD framework is outlined schematically in Figure 6.11. The model is divided into a number of submodels, each defined by a number of constraints as given by Equation (6.5). The submodels related to Materials, Energy, Technology, Recycling and Products are supported by LCA databases on a range of materials, energy, processes, waste management options, and products, respectively. In addition, the Products submodel also includes possible cascades of uses for a particular product and the constraints on the product, such as technical performance, legislative norms, *etc.* The submodel describing Supply Chain incorporates environmental and other data on the suppliers, while the model on Consumers contains data on consumer behaviour and preferences. Each option is analysed by the optimisation model, which then returns the optimum solutions for a particular product and the possible cascaded uses of a material or materials that make up the product. Thus, a plethora of Pareto optimum solutions is obtained for the decision-makers to elicit preferences for the options most suitable for a particular situation. Therefore, this approach enables forward planning and mapping of the materials and product flows in the economy and can ultimately lead to a more sustainable use of resources.

Key Facts ⚷
● By analysing all possible cascade options, LCPD enables forward planning and mapping of material flows, leading to a more sustainable use of resources.

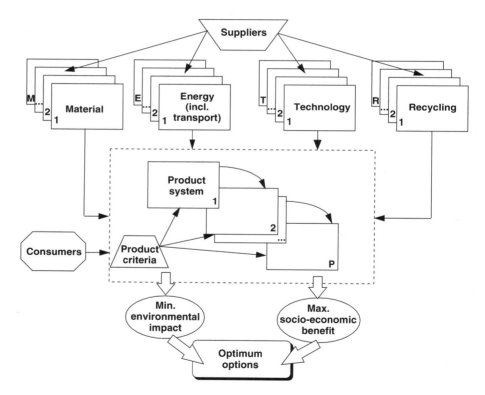

Figure 6.11 Mathematical modelling framework for LCPD

The submodels and their corresponding constraints and the objectives are described in more detail below.

Constraints

Each submodel is defined by material balance constraints. The environmental burdens and economic benefits are also calculated for each submodel. Furthermore, depending on their type, the submodels can have additional constraints. For instance, Materials can be described by constraints on physical and chemical characteristics and material availability. The Technology, Recycling and Products submodels incorporate constraints on technical performance. The Recycling submodel also has constraints on material contamination, as various additives may determine the type of the recycling option available for a particular material. The Product submodel, for example, also includes constraints on production capacities, market demand and various legislative limits. The mathematical formulation of the submodels is given below; it may be noted that these are only some of the constraints that each submodel could include and that, depending on a particular case, the number and type of the constraints may be different.

The following constraints can be defined:

Mass balances:
$$\sum_{i=1}^{I} a_{j,i}^{(m,e,t,r,p)} x_i^{(m,e,t,r,p)} = 0 \qquad j = 1, 2, \ldots, J \qquad (6.6)$$

Performance criteria:
$$\mu_g^{(m,e,t,r,p)} \geq c_g^{(m,e,t,r,p)} \qquad g = 1, 2, \ldots, G \qquad (6.7)$$

Material availability:
$$A^{(m)} \leq S^{(m)} \qquad (6.8)$$

Energy availability:
$$A^{(e)} \leq S^{(e)} \qquad (6.9)$$

Contamination (additives):
$$\alpha_n^{(r)} \leq d_n^{(r)} \qquad n = 1, 2, \ldots, N \qquad (6.10)$$

Market demand:
$$Q^{(p)} \leq D^{(p)} \qquad (6.11)$$

Process capacities:
$$\sum x_i^{(p)} \leq C_u^{(p)} \qquad (6.12)$$

Environmental legislation limits:
$$L_{1,k}^{(p)} \leq B_k^{(p)} \leq L_{2,k}^{(p)} \qquad k = 1, 2, \ldots, K \qquad (6.13)$$

Other legislation limits:
$$L_{1,b}^{(p)} \leq L_b^{(p)} \leq L_{2,b}^{(p)} \qquad b = 1, 2, \ldots, B \qquad (6.14)$$

Costs:
$$F^{(m,e,t,r,p)} \leq P^{(m,e,t,r,p)} \qquad (6.15)$$

where the superscripts m, e, t, r, and p correspond to the Materials, Energy (including Transport), Technology, Recycling and Product submodels. The variables in the above constraints are defined in the notation list at the end of the chapter.

Objective Functions

The system comprising the submodels defined by Equations (6.6)–(6.15) is then optimised on a number of environmental and economic objectives. As mentioned earlier, the environmental objectives are defined as burdens or impacts:

Min environmental burdens:
$$B_k^{(m,e,t,r,p)} = \sum_{i=1}^{I} bc_{k,i}^{(m,e,t,r,p)} x_i^{(m,e,t,r,p)} \qquad (6.16)$$

or

Min environmental impacts:
$$E_l^{(m,e,t,r,p)} = \sum_{k=1}^{K} ec_{l,k}^{(m,e,t,r,p)} B_k^{(m,e,t,r,p)} \qquad (6.17)$$

which is equivalent to Equations (6.1) and (6.2), respectively.

The economic objectives can be defined by costs or profit or by any other measure of financial benefit.

Max socio-economic benefit:
$$F^{(m,e,t,r,p)} = \sum_{i=i}^{I} f_i^{(m,e,t,r,p)} x_i^{(m,e,t,r,p)} \qquad (6.18)$$

Additional objectives can also be optimised, depending on the goal of optimisation and the priorities set by designers, engineers, policy-makers and so on. For instance, the goal may be to maximise the number of cascaded uses of the material or the number of uses in the same system while minimising the environmental impact and maximising the economic benefit. It may be noted that a smaller number of objectives reduces the computational burden and difficulties in interpreting the results. The latter arises from the number of Pareto optimum solutions which increases exponentially with the number of objectives.

This generic methodological framework for LCPD is now illustrated by an example of polymeric materials used for packaging.

6.3.2 LCPD for Polymers

As we have already highlighted several times, the production of polymers not only uses non-renewable resources, *i.e.* oil, but it also results in an energy loss. Given the ever increasing rates of use of polymers, and the fact that the majority of these end up in a landfill, it is clear that reusing and recycling polymers would not only save non-renewable resource and reduce the amount of solid waste, but could also generate income. Thus, recycling polymers makes sense both environmentally and financially. The following example outlines how the LCPD methodology described above can be used to identify the optimum options for the re-use and recycling of polymers.

Figure 6.12 shows a life cycle of a polymer, a number of recycling options in that life cycle, and subsequent cascades of uses of the polymeric material. In addition to recycling in the same system, a polymer can also be cascaded to a different life cycle, for the manufacture of another product. This cascading can occur at any point in the recycling loop, depending on a number of criteria, as discussed above. The problem is, therefore, to determine the optimum point at which a termination of one life cycle should occur for a material to enter a new cascade. In terms of LCPD, a number of different polymers will normally be considered for a particular product and their recycling and cascading options examined for associated environmental and economic benefits. Furthermore, different technologies and energy delivery also play an important role, as do the supplier chain and the consumers. All

Key Facts
● The number of optimisation objectives depends on the goal and scope of the analysis.
● The number of Pareto optimum solutions increases exponentially with the number of objectives.

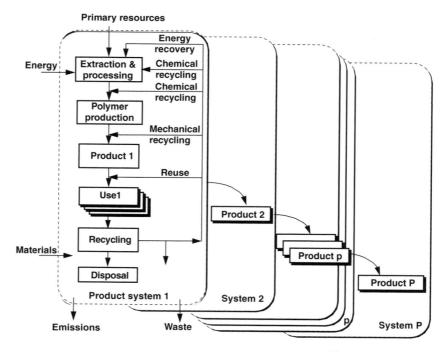

Figure 6.12 Cascaded use of polymeric materials[13]

these criteria are incorporated into the model, as discussed in the previous section, to give solutions for the optimum use of resources.

LCPD Modelling: An Example of HDPE Bottles

To illustrate how the LCPD methodology can be applied to the design of polymer products and in particular to their cascaded use, the example of HDPE bottles discussed in Section 6.2.2 is further developed below. From their first life cycle, the HDPE bottles are cascaded to become a plastic crate in their second, and a carpet lining in the third and final life cycle[13,14].

Figure 6.13 outlines the system comprising the three life cycles. To simplify the explanation, a small number of constraints is considered and only the material, energy (including transport) and product subsystems are modelled; the technology, recycling, and supplier options are given and fixed.

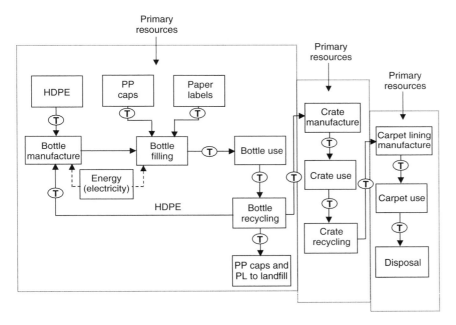

Figure 6.13 An example of cascaded use of HDPE bottles (T-transport)

First Life Cycle

As in the example in Section 6.2.2, the bottles are made up of HDPE, PP caps and paper labels (PL). The 'cradle to gate' life cycles of HDPE and PP, from extraction of raw materials to polymer production, can be found in Figure 6.6. The functional unit is defined as packaging 1000 l of water in 1-l bottles. The quantities of HDPE, PP caps and PL entering the use phase are 47 kg, 2 kg and 1 kg, respectively. After use, the empty bottles are collected and sent to a recycling point where the labels and caps are separated from the bottles and sent to landfill. The bottles are sent for reprocessing; they are blended with virgin material before being used again for bottle production. Thus this system is almost identical to that in the LCA example in Section 6.2.2; the only difference is that HDPE is no longer incinerated but is sent for recycling (compare Figures 6.5 and 6.13).

CONSTRAINTS

The quantities of materials in the bottle system are as follows:

- the mass of HDPE: $x_1 = 47$ kg
- the mass of PP caps: $x_2 = 2$ kg

- the mass of paper labels: $x_3 = 1$ kg
- the total mass of HDPE bottles: $x_4 = 50$ kg

Following the LCPD methodology, a simplified overall material balance in the bottle system as defined by Equation (6.6) is:

$$\text{Mass balance:} \quad x_1 + x_2 + x_3 - x_4 = 0 \tag{6.6'}$$

Each of these materials is characterised by a set of properties or performance criteria that must be met if they are to be used in this system. The performance constraints in this example are defined as follows:

- quantity of each material for recycling: $\mu_1 \geq 5$ kg
- optical transparency of the HDPE bottle: $\mu_2 \geq 75\%$
- haze of the HDPE bottle: $\mu_3 \leq 0.8\%$
- tensile strength of the PP cap: $\mu_4 \geq 25$ MPa
- tear strength of the label: $\mu_5 \geq 20$ N mm^{-1}

According to Equation (6.7), the performance constraints are thus formulated as:

Performance criteria:

HDPE: $\qquad\qquad \mu_1^{(m_1)} \geq 5; \quad \mu_2^{(m_1)} > 0.75; \qquad \mu_3^{(m_1)} \leq 0.008;$

PP cap: $\qquad\qquad \mu_1^{(m_2)} \geq 5; \quad \mu_4^{(m_2)} \geq 25;$ $\qquad\qquad$ (6.7')

PL: $\qquad\qquad\quad \mu_1^{(m_3)} \geq 5; \quad \mu_5^{(m_3)} \geq 20.$

In addition, one environmental legislation constraint has been placed on the system. It is assumed that total NO$_x$ emissions for the blow moulding of 1000 bottles must not exceed a compliance limit of 120 g. According to Equation (6.13), this constraint is defined as:

$$\text{Environmental legislation:} \quad B_k^{(p)} \leq 120 \tag{6.13'}$$

Finally, there is an economic constraint on the system: the market price P of the bottle must be greater than the total cost of production F. In this example, the market price is taken as £300 per 1000 bottles. Thus, according to Equation (6.15) we have:

$$\text{Market constraint:} \quad F < 300 \tag{6.15'}$$

OBJECTIVE FUNCTIONS

For illustration, only one environmental impact (acidification) is considered. It is defined using the formulation (6.17):

$$\text{Min} \quad E = \Sigma ec_k^{(m,e,p)} B_k^{(m,e,p)} \tag{6.17'}$$

which is, in effect, equivalent to Equation (6.2) and the way in which the impacts were calculated in the LCA example in Section 6.2.2. Since there is only one impact, the subscript l is dropped for simplified notation. As noted earlier, the impacts from the energy submodel e here also include transport.

Two types of costs considered are materials (f_i^m) and energy (f_i^e), including transport. As defined by Equation (6.18), economic costs accumulated over the life cycle of 1000 HDPE bottles are:

$$\text{Min} \quad F = \Sigma f_i^{(m)} x_i^{(m)} + \Sigma f_i^{(e)} x_i^{(e)} \tag{6.18'}$$

It is also assumed to be uneconomic to recover and recycle less than 5 kg of any material per functional unit (defined in Equation (6.7') as a performance criterion constraint, μ_1).

As in the LCA example illustrating polymeric packaging (Section 6.2.2), this example also assumes that, in the first life cycle, the 1000 bottles are produced solely from virgin materials, so that the ratio of virgin material to recyclate is 100:0. The input properties of the three materials into the bottle manufacture are:

- optical transparency of the HDPE bottle: $\mu_2 = 94\%$
- haze of the HDPE bottle: $\mu_3 = 0.8\%$
- tensile strength of the PP cap: $\mu_4 = 30\,\text{MPa}$
- tear strength of the label: $\mu_5 = 22\,\text{N mm}^{-1}$

To ensure the materials are suitable, the model compares the material properties with the process requirements as defined by Equation (6.7'). The properties of the three material inputs are within the performance constraints, and they are consequently deemed fit for purpose.

Changes in the properties of a polymer can occur as a result of many different processing and environmental effects (see Chapter 2). In this example, it is assumed that sufficient anti-oxidant has been added to the HDPE to prevent degradation during the moulding of the bottle and cap. It is also assumed that because of the relatively short shelf-life of the products (several months), no significant UV degradation would occur. As a consequence, the performance parameters remain constant throughout use.

Table 6.2 shows the results of the LCPD model obtained by optimising on the environmental impact (acidification) defined by Equation (6.17'). The table lists the NO_x formation from each activity and the associated acidification potential. Total (minimised) life cycle NO_x emissions from the production of 1000 water bottles from virgin HDPE are therefore 1.045 kg, which is, in this example, equivalent to an acidification potential of 1.35 kg $SO_{2\,\text{equiv}}$. Note that this differs from the LCA result for polymeric packaging in Section 6.2.2 because there is no incineration in this example. It is clear from the table that the optimisation results satisfy the constraint on NO_x emissions from water bottle production, which creates 115 g of NO_x. That is within the compliance limit of 120 g of NO_x (Equation 6.13').

Table 6.2 Environmental burdens and impacts for the HDPE bottle system: the first life cycle.

Activity	NO_x (kg per 1000 bottles)	Acidification (kg $SO_{2\,\text{equiv}}$ per 1000 bottles)
HDPE production	0.365	0.473
PP caps	0.022	0.028
Electricity	0.080	0.104
Paper labels	0.003	0.004
Transport	0.460	0.596
Production of water bottle	0.115	0.149
Total	1.045	1.350

Table 6.3 summarises the costs associated with the activities in the system, up to the water bottle manufacture. These results have been obtained by optimising on the cost objective function given by Equation (6.18'). Total cost for materials, transport and energy is £89.22. Thus, this satisfies the economic constraint (Equation 6.15') that the total production cost must be less than the market price of the bottles, taken as £300.

Cascaded Use of Materials: Further Lives

If the materials are to be recovered, the constraints in force must continue to be met. Maximum efficiency for this activity would result in the mass flows of the label, cap, and bottle being 1 kg, 2 kg and 47 kg, respectively. The constraint which states that 5 kg of

Table 6.3 Life cycle economic costs of the HDPE bottle system: the first life cycle.

Activity	Costs (£)	Cumulative costs (£)
HDPE production	69.90	69.90
PP caps	3.00	72.90
Electricity	10.00	82.90
Paper labels	0.60	83.51
Transport	1.41	84.91
Production of water bottle	4.31	89.22

material is necessary for recycling to be economic means that the cap and paper are not suitable for recycling in this example. Only the HDPE meets the minimum flow recycling requirements and is allowed by the LCPD model to proceed to the recycling point.

It is assumed that the efficiency of the recycling activity is such that out of every 1000 bottles processed, 10 bottles pass through the activity unchanged. The associated PP and paper represent contamination of the HDPE to the level of 0.064 % w/w. This, and the other performance parameters for the recovered HDPE, are listed in Table 6.4.

Table 6.4 Performance parameters for the recycled HDPE bottles.

Activity	Haze (%)	Contamination level (%)	Optical transparency (%)
After use	0.600	0	93.75
Recycling/reprocessing	0.856	0.064	69.75
Blending with virgin HDPE	0.664	0.016	87.75

Recycling activity granulates the HDPE so that it is in a form that can easily be blended with virgin material. It is assumed here that the anti-oxidant used for original bottle processing is sufficient to stop any further material degradation and therefore the performance parameters remain constant throughout this process.

Water bottle manufacturers will only accept material whose performance properties meet the criteria defined in Equation (6.7'). Contamination can affect haze and optical

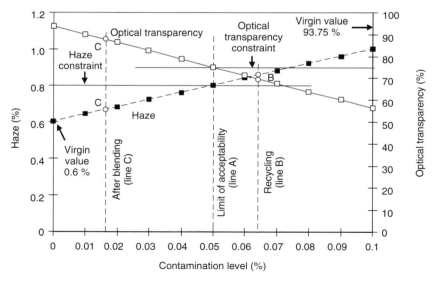

Figure 6.14 Performance parameter change in the HDPE bottle system[14]. Reproduced with permission from Mellor, W.; Williams, E.; Clift, R.; Azpagic, A. and Stevens, G.C. (2002). 'A framework for lifecycle product design (LCPD).' In *Cascaded Systems, Chem. Eng. Sci.* Currently in press. Copyright, Elsevier Science

transparency. The changes in performance parameters are illustrated in Figure 6.14. It is assumed that a contamination level of 0.05 % results in haze of 0.8 % and optical transparency of 75 %, at the limits of acceptability (Line A); to the right of this line the criteria do not satisfy the constraints. It is also assumed that an increase of 0.1 % in the contamination level will result in a 5 % increase in haze and a 5 % reduction in optical transparency.

The contamination level of 0.064 % (Line B) therefore results in haze being increased to 0.856 % and the optical transparency reduced to 69.75 % (points B). The values of these utilities are beyond the limit of acceptability and, therefore, mean that in its current state the polymer does not meet the requirements for water bottle manufacture.

In order to reduce the level of contamination and thus the haze and optical transparency of the polymer, blending of the recyclate with virgin polymer is carried out in the ratio of 75 % virgin to 25 % recyclate. At this ratio (Line C), the haze and optical transparency values become 0.664 % and 87.75 %, respectively (points C), which are within the performance constraints. Consequently, with appropriate blending the polymer can be suitable for re-use in the same application.

However, if for some reason the appropriate blending conditions cannot be achieved, the material can be passed on to the next cascades of use, in our case for manufacturing crates and carpet lining. An additional set of constraints would have to be defined to reflect the requirements of these cascades of use, including the performance and cost constraints. Optimisation on economic and environmental objectives would then be performed on the whole system to give a set of Pareto optimum solutions. This would then enable decision-makers to map the flows of the HDPE material from one life cycle to another and to identify the sustainable paths that minimise environmental impacts and maximise socio-economic benefit.

It is worth noting that the point at which a material leaves one life cycle and enters another will depend on a number of constraints and objectives, so that for different optimisation criteria the optimum solutions will be different. It is therefore incorrect to assume that, for every material and product, recycling for the same application is a better option than cascading it into a different use. The right conclusions can only be arrived at by considering all options simultaneously and comparing their environmental and economic implications, as illustrated by the simple example of HDPE.

> **Key Facts**
> ● When a material can undergo a number of cascades of use and re-use, the point at which is passes from one cascade to the next depends on the number of constraints and objectives.
> ● It is not always best to recycle within a single application, sometimes the environmental or economic gains are greater if the polymer is cascaded to the next application.

6.4 SUMMARY AND LEARNING OUTCOMES

The move towards sustainable development requires a paradigm shift from a fractured view of the environment, with the emphasis on one stage of the life cycle, to a more holistic life cycle approach to environmental management. Life Cycle Assessment is a tool that enables and supports such a paradigm shift as it embodies life cycle thinking and so provides a full picture of human interactions with the environment.

An emerging application of life cycle thinking and LCA is in design for the environment. A developing Life Cycle Product Design (LCPD) tool, which combines LCA and system optimisation, offers a potential for technological innovation in the polymer products concept and structure through selection of the best material and process alternatives over the whole cycle. It also enables tracking and optimising of the flows of polymer materials and products through the economy with the aim of identifying sustainable ecologies of materials.

After studying the material presented in this chapter, you should be able to understand and discuss the following:

- the advantages and disadvantages of adopting the life cycle approach to analyse economic systems;
- the role of LCA in identifying sustainable options, particularly in the context of resource and waste management;
- the methodology of LCA and LCPD;
- the relevance and application of LCA and LCPD to polymers;
- the life cycle implications of different resource and waste management strategies.

Notation

A	Availability of a resource or technology
A_j	Right-hand side coefficient j in constraint j
$a_{i,i}$	Input/output coefficient j for activity i
B_k	Environmental burden k
$Bc_{k,i}$	Environmental burden coefficient k for activity i
C_i	Capacity of process or activity x_i
c_g	Minimum requirement on performance criterion g
D	Market demand for product p
d_n	Minimum concentration of additive n
E_l	Environmental impact l
$ec_{l,k}$	Environmental impact coefficient l for burden k
F	Economic costs
f_i	Economic coefficient i related to activity x_i
$L_{1,b}$	Other legislation lower limit b
$L_{2,b}$	Other legislation upper limit b
$L_{1,k}$	Environmental legislation lower limit for burden k
$L_{2,k}$	Environmental legislation upper limit for burden k
P	Economic profit or price
Q	Production (quantity) of product p
S	Supply of a resource or technology
x_i	Activity or operation level of process
α_n	Additive or contaminant n in a material
μ_g	Performance criterion g

Superscripts

e	Energy (including transport)
m	Materials
p	Products
r	Recycling options (and technologies)
t	Technologies

6.5 REFERENCES AND FURTHER READING

1. Daae, E. and Clift, R. (1994). A life cycle assessment of implications of paper use and recycling, *IChemE Environ. Protect. Bull.*, **28**, 23–25.
2. McQuaid, J. (Ed.) (1995). *Engineering for Sustainable Development*, The Royal Academy of Engineering, London, p. 160.
3. ISO (1997). ISO/DIS 14040: *Environmental Management – Life Cycle Assessment – Principles and Framework*, HMSO, London.
4. ISO (1998). ISO/DIS 14041: *Environmental Management – Life Cycle Assessment – Goal and Scope Definition and Life Cycle Inventory Analysis*, HMSO, London.
5. ISO (1998). ISO/CD 14042: *Environmental Management – Life Cycle Assessment – Life Cycle Impact Assessment*, HMSO, London.
6. ISO (1998). ISO/CD 14043: *Environmental Management – Life Cycle Assessment – Life Cycle Interpretation*, HMSO, London.
7. Heijungs, R. *et al.* (eds) (1992). *Environmental Life Cycle Assessment of Products: Background and Guide*, MultiCopy, Leiden.
8. Ecobilan (1995–2001). TEAM LCA software and DEAM database, PWC, London
9. PIRA (1996–2001). PEMS LCA software and database, PIRA, Leatherhead, UK.
10. PRé Consultants (1995–2001). SimaPro LCA software and database, The Netherlands.
11. Alting, L. (1996). Life cycle engineering and design, Proc. of Clean Technology Conference, Royal Society, London.

12. Azapagic, A. (1997). Life Cycle Assessment: A tool for innovation and improved environmental performance. In *Science, Technology and Innovation Policy: Opportunities and Challenges for the Knowledge Economy*, ed. Conceicao, P., Gibson, D., Heitor, M.V. and Shariq, S. Part VI.35, Quorum Books, Westport, CT, pp. 519–530.
13. Azapagic, A. (1998). Design for optimum use of resources – cascaded use of materials. In *Knowledge for Inclusive Development*, ed. Conceicao, P., Gibson, D., Heitor, M.V., Shariq, S., Sirilli, G. and Veloso, F. Edward Elgar Publishers, Cheltenham, UK, Chapter 20, in press.
14. Mellor, W.E., Williams, E., Clift, R., Azapagic, A. and Stevens, G. (2002). A framework for life cycle product design (LCPD) in cascaded systems, *Chem. Eng. Sci.*, in press.
15. Mellor, W., Williams, E.A., Clift, R., Azapagic, A. and Stevens, G.C. (2001). A mathematical model and decision-support framework for material recovery, recycling and cascaded use, *Chem. Eng. Sci.*, in press.
16. Azapagic, A. (1996). Environmental system analysis: the application of linear programming to life cycle assessment. PhD Dissertation, University of Surrey.

Further Reading

Andersson, K. and Ohlsson, T. (1999). Life cycle assessment of bread produced on different scales, *Int. J. LCA*, **4**, 25–40.

APME (1992–1997). Reports 1–13. LCA studies of various plastics. APME, Brussels. Also available at: http://www.apme.org.

Arentsen, H., Van Lochem and Van Steenderen (1992). In *Polymer Products and Waste Management: A Multidisciplinary Approach*, ed. Smits, M., International Books, Utrecht.

Aresta, M. and Tommasi, I. (1997). Carbon dioxide utilisation in the chemical industry, *Energy Convers. Manage.*, **38**(SS), S373–S378.

Audsley, E. *et al.* (1997). *Harmonisation of Environmental Life Cycle Assessment for Agriculture*. Final Report. Concerted Action AIR3-CT94-2028, Silsoe, UK.

Azapagic, A. (1999). Life cycle assessment and its application to process selection, design and optimisation, *Chem. Eng. J.*, **73**, 1–21.

Azapagic, A. and Clift, R. (1999a). Allocation of environmental burdens in multiple-function systems, *J. Cleaner Prod.*, **7**, 101–119.

Azapagic, A. and Clift, R. (1999b). Allocation of environmental burdens in co-product systems: product-related burdens (part 1). *J. Int. LCA*, **4**(6), 357–369.

Azapagic, A. and Clift, R. (1999c). Life cycle assessment as a tool for improving process performance: a case study on boron products, *Int. J. LCA*, **4**(3), 133–142.

Azapagic, A. and Clift, R. (1999d). The application of life cycle assessment to process optimisation, *Comput. Chem. Eng.*, **23**, 1509–1526.

Azapagic, A. and Clift, R. (1999e). Life cycle assessment and multiobjective optimisation, *J. Cleaner Prod.*, **7**(2), 135–143.

Azapagic, A. and Clift, R. (2000). Allocation of environmental burdens in co-product systems: process and product-related burdens (part 2). *J. Int. LCA*, **5**(1) 31–36.

Backlund, B. (1998). Scandinavian collaboration develops life cycle assessment (LCA) as a tool for the forest industry, *Svensk Papperstidning-Nordisk Cellulosa*, **101**(3), 49–50 (in Swedish).

Beck, A., Scheringer, M. and Hungerbühler, K. (2000). Fate modelling within LCA, *Int. J. LCA*, **5**(6), 335–344.

Bell, G.M., Azapagic, A., Faraday, D.B.F. and Schulz, R.A. (2000). Sustainable practices for potable spirits manufacturing: traditional *vs.* alternative processes, AIChE Spring Meeting, Atlanta, 5–9 March 2000.

Boustead, I. (1992). *Eco-balance Methodology for Commodity Thermoplastics*, PWMI, Brussels.

Cederberg, C. (1998). Life cycle assessment of milk production – a comparison of conventional and organic farming, SIK Report No. 643, The Swedish Institute for Food and Biotechnology (SIK), Gothenburg, Sweden.

Chubbs, S.T. and Steiner, B.A. (1998). Life cycle assessment in the steel industry, *Environ. Prog.*, **17**(2), 92–95.

Consoli, F., Allen, D., Boustead, I., Fava, J., Franklin, W., Jensen, A.A., de Oude, N., Parrish, R., Perriman, R., Postlethwaite, D., Quay, B., Séguin, J. and Vigon, B. (eds) (1993). *Guidelines for Life-Cycle Assessment: A 'Code of Practice'*, SETAC, Brussels.

de Langhe, P., Criel, S. and Ceuterick, D. (1998). Green design of telecom products: the ADSI, high speed modem as a case study, *YEE Trans. Components Packag. Manuf. Technology, Part A*, **21**(1), 154–167.

Dobson, I.D. (1996). Life cycle assessment for painting processes: putting the VOC Issue in perspective, *Prog. Org. Coatings*, **27**(14), 55–58.

Fava, J., Consoli, F., Denison, R., Dickson, K., Mohin, T. and Vigon, B. (eds) (1993). *A Conceptual Framework for Life-Cycle Impact Assessment*, SETAC and SETAC Foundation for Environmental Education, Inc., Pensacola.

Franke, M., Kluppel, H., Kirchert, K. and Olschewski, P. (1995). Life-cycle assessment – life-cycle inventory for detergent manufacturing, *Tenside Surfact. Deterg.*, **32**(6), 508–514.

Graedel, T.E. and Alenby B.R. (1995). *Industrial Ecology*, Prentice Hall, Englewood Cliffs, NJ.

Haas, G., Wetterich, F. and Geier, U. (2000). Life cycle assessment framework in agriculture on the farm level, *Int. J. LCA*, **5**(6), 345–348.

Kuusinen, T.L., Barker, R.H. and Alexander, D.A. (1998). Life cycle assessment in woven textiles, *Tappi J.*, **81**(3), 179–182.

Miyamoto, S. and Tekawa, M. (1998). Development of life cycle assessment software and application to personal computer assessment, *NEC Res. Dev.*, **39**(2), 77–81.

Ophus, E. and Digernes, V. (1996). Life-cycle assessment of an alkyd emulsion: improvements in environmental performance, *Jocca-Surf. Coat. Int.*, **79**(4), 156.

Puntener, A.G. (1998). Risk assessment of leather dyestuffs, *J. Soc. Leather Technol. Chem.*, **82**(1), 1–4.

Rice, G. (1997). Life cycle assessment of carbon dioxide production. In: *The Application of Life Cycle Assessment to Industrial Process Selection*, EngD Portfolio, University of Surrey.

Robertson, J.G.S., Wood, J.R., Ralph, B. and Fenn, R. (1997). Analysis of lead/acid battery life cycle factors: their impact on society and the lead industry, *J. Power Sources*, **67**(1–2), 225–236.

Seppala, J., Melanen, M., Jouttijarvi, T., Kauppi, L. and Leikola, N. (1998). Forest industry and the environment: a life cycle assessment study from Finland. *Resour. Conserv. Recycling*, **23**(1–2), 87–105.

Yoda, N. (1996). Life cycle assessment in polymer industry toward the 21st century. *J. Macromol. Sci. – Pure Appl. Chem.*, **A33**(12), 1807–1824.

6.6 REVISION EXERCISES

1. Why is the life cycle approach fundamental to identifying more sustainable solutions?

2. Describe the methodology for LCA, as defined by ISO 14040.

3. Why is allocation a problem in LCA? Give examples of different allocation approaches.

4. How do you think the system boundary affects the complexity of an LCA study and the results?

5. How does the data quality affect the results of an LCA study?

6. How do you think LCA software and databases can help in conducting an LCA?

7. Is there a potential for misuse of LCA results and if so, by whom?

8. Consider the LCA example in Section 6.2.1. Why do you think PP is better in environmental terms than HDPE for some impacts but worse for the others?

9. Describe a life cycle of polyurethane from 'cradle to gate'. Draw a flow diagram to represent its life cycle. How would the life cycle look if you were to include a few different uses of PU? Draw the flow diagrams to describe the life cycles from 'cradle to grave' for different uses.

10. Why is it important to define the functional unit correctly? Discuss this with reference to the following examples:

   - packaging materials,
   - wall paints,
   - floor coverings,
   - transport.

   How would you define the functional units for these systems, to be able to identify the most sustainable alternative?

11. Describe the methodology for Life Cycle Product Design (LCPD). In your opinion, what is the difference between Life Cycle Product and Process Design? Support your explanation with reference to relevant examples.

12. How can LCPD be applied to polymers?

13. What is the Pareto optimum? How is it relevant to decision-making?

14. Who, in your opinion, should be involved in the decision-making to identify more sustainable options for polymers? Why?

15. Examine the kettle in your house: how would you go about designing a more sustainable one? Justify your decisions regarding the material and energy use along the whole life cycle. Which stage in the life cycle do you think contributes to environmental impacts the most? How would you reduce the impacts from that stage?

environmental impacts of recycling

Chapter 7 – The Mirror of Venus (E Burne-Jones, 1870 – 76)
Burne-Jones painted this painting without reference to a specific myth: Venus and her nubile handmaidens survey their own reflections in a limpid pool, its surface covered by the broad leaves of water-lilies (incidentally the title of one of the examples examined in detail in this chapter).

7.1 INTRODUCTION

As we have seen in Chapters 1 and 4, recycling options for polymeric materials comprise mechanical and chemical recycling and incineration with energy recovery. All these options have certain advantages and disadvantages. For example, mechanical recycling involves relatively simple technologies but, owing to the need to identify and separate the individual plastics, can be quite labour- or energy-intensive, depending on whether the separation process is manual or automated. The primary advantage of chemical recycling is that it enables recycling of mixed or soiled waste plastic; however, it can have high capital costs. Both mechanical and chemical recycling preserve the non-renewable carbon-based resources that are locked up in polymeric materials. This may therefore favour them over incineration with energy recovery which, despite the resulting energy generation, may be viewed as a waste of finite non-renewable resources.

In addition to these and other technological and economic advantages or disadvantages (see Chapters 4 and 5 for more detail), different recycling options and technologies must also be assessed and compared on the basis of environmental performance. Although it is often assumed that recycling of plastics is more sustainable than their disposal to landfill after being used only once, viewed on a life cycle basis, it is clear that recycling is not totally impact-free. Like any other industrial process or technology, it uses energy and materials and generates additional air and water emissions and solid waste. Furthermore, as we have already seen in Chapter 5, it usually involves complex reverse logistics associated with the recovery of waste plastics from consumers, often resulting in a large number of transport steps, which require the use of fossil fuels and generate air emissions.

Thus, it is important to realise that it cannot be considered environmentally sustainable to recycle if the process uses more resources and energy than can be gained by recycling. In this chapter we examine the life cycle environmental profiles of different recycling technologies to identify more sustainable options. Following a general discussion in the next section, we will examine in greater detail the specific advantages and disadvantages of recycling by considering several real LCA case studies, including recycling of plastic packaging, car windscreen polymer interlayers, furniture cushioning and plastic panels used for electronic equipment.

7.2 ENVIRONMENTAL IMPACTS OF RECYCLING: LIFE CYCLE CONSIDERATIONS

Figure 7.1 shows a generic life cycle flow diagram of the four end-of-life options for waste plastics considered in this chapter: re-use, mechanical and chemical recycling (fuel or monomer production) and incineration with energy recovery. Simplified life cycles of each option are depicted in Figure 7.2. After use, plastic waste has to be collected from domestic or commercial users and transported to a recycling point.

Direct re-use of plastic products (usually in the same, primary, application) requires collection of waste and some refurbishment or remanufacturing. For example, when electrical or electronic equipment reaches the end of its useful life, plastic parts can be separated from other materials, refurbished to repair any damage, repainted if necessary and re-used on new equipment. Each of these activities requires additional energy (e.g. transport) and materials (e.g. painting) and generates emission to air, water and land. We will discuss these aspects in more detail in Section 7.3.1 by examining a case study of plastic panels mounted on photocopying machines.

As explained in Chapter 4 and shown in Figure 7.2, mechanical recycling involves waste collection and transportation, sorting to separate individual plastic materials and grinding or re-melting to produce plastic pellets. We have already noted that, depending on whether the sorting process is automatic or manual, this stage can be energy- or labour-intensive, thus contributing additional environmental burdens. Grinding and remelting also require energy input; for example electricity consumption to form recycled bottle granulate is

Key Facts 🔑
- Recycling is not itself free of environmental impacts because it uses materials and energy and produces air and water emissions and solid waste.
- Transportation of material for recycling also has a negative impacts through use of non-renewable fuels and generation of air emissions.

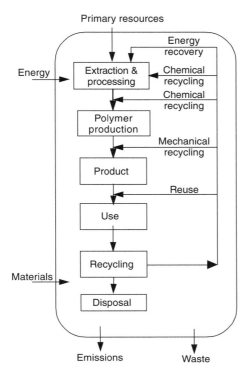

Figure 7.1 Life cycle diagram of different polymer recycling options

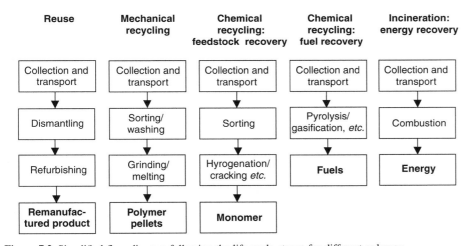

Figure 7.2 Simplified flow diagram following the life cycle stages for different polymer recycling options

420 MJ *per* 100 kg[1]. This figure represents 14% of the total energy of 3000 MJ *per* 100 kg consumed in the life cycle of bottle manufacturing (see Figure 6.7 in Chapter 6).

The process of chemical recycling is shown in Figure 7.2 and follows the life cycles of feedstock and fuel recovery. In feedstock recycling, after collection, transportation and sorting, the waste is depolymerised, for example, in a hydrogenation or hydrocracking process to produce lower molecular weight products, which can then be used as a raw material to make new polymers. Fuel recovery from waste plastics has similar steps, except that the technologies used here may be different from those for feedstock recycling. For instance, gasification and pyrolysis are often quoted as the most suitable candidate technologies for fuel recovery, producing synthesis gas (CO and/or H_2) and oil and gas, respectively. You may remember that chemical recycling technologies were discussed in more detail in Chapter 4.

The final recycling option shown in Figure 7.2 is incineration with energy recovery. This option does not require major sorting, although elimination of PVC (discussed in more detail in Chapter 4) from the mixed waste stream will minimise emissions of dioxins and hydrochloric acid. The combustion stage comprises a furnace, a system for heat recovery from the flue gas and a system for flue gas treatment (see Figure 4.12 in Chapter 4). Depending on the type of combustor used, it may be necessary to shred the waste first in order to reduce it to smaller fragments. This is the case in fluidised-bed combustion. Grate firing on the other hand does not require pretreatment of plastics prior to incineration. The steam that may be recovered from incineration can be used to generate both electricity and heat, thus displacing the need for a dedicated fossil-fuel based power plant. However, this option is not without its disadvantages: unlike other recycling options, it depletes the non-renewable resources and generates both air emissions and solid waste. For example, CO_2 emissions from combustion of most commodity plastics range from 3100 to 3400 kg of CO_2 *per* tonne of plastics (except for PVC, which has a low carbon content)[2]. By comparison, LCA studies show that, on average, 1500–2000 kg of CO_2 is produced in the life cycle of plastics products[3]. For the life cycle of plastic bottles, the average CO_2 emissions range from 1800–3500 kg tonne^{-1} (see Figure 6.7 in Chapter 6; note that CO_2 emissions are expressed *per* functional unit, not *per* tonne).

For the remainder of this chapter it will be interesting to examine in more detail how the life cycles of these recovery options compare for different plastic products and materials. For that we have chosen four genuine LCA case studies, each representing a different, generic set of options. The first three studies examine recycling possibilities in cases where the focus is on the end-of-life management, comprising closed-loop recycling (plastic panels), cascaded use (laminated car windscreens) and integrated waste management (packaging). In the final case study, the emphasis is shifted to the front end of a product's life: its design. This study shows, through the example of a 'Waterlily' mattress, the benefits of the design for chemical recycling.

7.3 LCA CASE STUDIES OF RECYCLING OPTIONS AND TECHNOLOGIES

7.3.1 Closed-Loop Recycling: Plastic Panels

The plastic panels that are mounted on various items of office electronic equipment, such as photocopiers, computers, telephones and fax machines, are mainly used for aesthetic purposes, giving the equipment a final, hopefully pleasing, shape. Given the size of this market and the ever-shortening life cycles of the electronic equipment, it is becoming increasingly important to identify sustainable end-of-life options for plastic panels, if disposal of large amounts of solid waste in landfills is to be avoided.

This is the aim of the first LCA case study that we will examine. This particular study[4,5] was initiated by an office equipment manufacturer. At present, the majority of panels are refurbished; however, a number of constraints limit their refurbishment and the company is considering other options to reduce the amount of solid waste that is sent to landfill.

For these purposes, the environmental impacts of refurbishment of the rear plastic panels used on photocopying machines are compared with mechanical recycling of the polymer for the manufacture of new panels. Two other end-of-life options are also considered: incineration of the polymer (without energy recovery) and landfilling.

The life cycle diagram of the system under study is shown in Figure 7.3. The functional unit (see Chapter 6 for the definition of a functional unit) is defined as the manufacture and use of 19 000 panels, which is equivalent to the annual panel demand in the UK. The plastics used in the production of panels are engineering grades of polycarbonate (PC) and

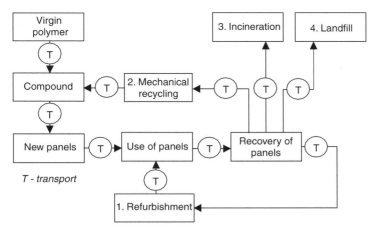

Figure 7.3 Life cycle flow diagram illustrating the production, recovery and recycling options for plastic panels

Within the figure:
- Virgin polymer → T → Compound
- Compound → T → New panels
- New panels → T → Use of panels
- Use of panels → T → Recovery of panels
- 2. Mechanical recycling → T → Compound
- 1. Refurbishment → T → Use of panels
- 3. Incineration
- 4. Landfill
- *T - transport*

poly(acrylonitrile-*co*-butadiene-*co*-styrene) (ABS) (you may recall that these polymers were discussed in Chapters 2 and 3).

After the use phase, the photocopiers are recovered and the panels dismantled to be refurbished, recycled, incinerated and/or landfilled. Refurbishment involves panel cleaning and repair (in which scratches and other repairable damage are filled and sanded). The panels are then re-sprayed, usually using water-based paints (we will discuss these paints again in Chapter 8). As noted earlier, there are limits to the extent to which panels can be remanufactured. Technological and design change of the equipment, high aesthetic standards, brand-specific components and logistics complexity are all reasons why a new panel may be a preferred option rather than refurbishment. Unfortunately, a panel can only be refurbished once, due to problems associated with unsatisfactory re-painting.

Mechanical recycling breaks down the plastic panels mechanically to yield granulate. This recyclate can easily be compounded with the virgin material. Only panels without surface treatment can currently be mechanically recycled. This means that refurbished panels cannot be recycled and implies that the panels can only be used once if the plastic material is going to be reprocessed.

The study has considered a number of scenarios, as illustrated in Figure 7.4. In scenario A, 19 000 panels are made from virgin polymer. The panels are assembled, used once and eventually dismantled and sent to landfill. In scenario B, 10 000 panels are manufactured from virgin polymer. Together with 9000 refurbished panels, they are assembled into 19 000 photocopiers. After use, the panels are dismantled and are routed as follows: 9000 panels used only once ('new' panels) are refurbished, the remaining 1000 'new' panels are sent to landfill, and all 9000 refurbished panels are also sent to the landfill.

Scenario C considers a combination of refurbishment, recycling and incineration. Here, 9000 panels are made from virgin polymer. In addition, 9000 'new' panels are refurbished after the first use and 1000 panels (also after the first use) are recycled mechanically to be blended with virgin polymer (owing to quality constraints, only 25 % of recycled material can be blended with virgin polymer) and make 1000 brand new panels. The remaining 9000 refurbished panels are incinerated after their now second use. Scenario D is the same as scenario C, except that the panels are not incinerated but sent to landfill.

Finally, in scenario E, 12 600 panels are manufactured from virgin polymer and 6400 panels from a combination of virgin and recycled polymer (again, in the ratio 75 % virgin and 25 % recycled material). After use, 12 600 recycled panels are disposed in a landfill site.

The environmental impacts of these different scenarios are shown in Figure 7.5 (see Appendix 2 for the definition of impacts). The results show that in this particular case study Scenario D, which combines refurbishment, recycling and landfilling of panels, is

Key Facts 🗝
- Issues such as aesthetic design changes and logistics limit the use of refurbished plastic components.
- Surface treatments (*e.g.* painting) limit opportunities for mechanical recycling after components have been refurbished.

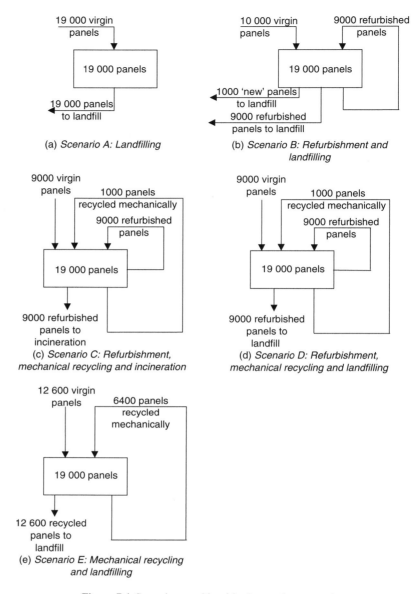

Figure 7.4 Scenarios considered in the panels case study

environmentally the most sustainable option. The exception to this is the amount of waste that is sent to landfill for which scenario C is the best option. However, this scenario is not sustainable with respect to other impact categories, particularly for aquatic ecotoxicity where there is a nine-fold increase compared to option D. It should be noted, however, that this case study assumes that no energy is recovered in the incineration of plastics. If energy recovery were included, then the system would have to be credited for the avoided burdens (see Chapter 6) and this might alter the overall conclusions of the study.

Option B, in which a proportion of panels is refurbished and then landfilled, is the second preferred option. The difference in impacts between scenarios B and D is on average 4 %. Scenario B also happens to be the end-of-life option currently practised by the company. Scenario A, in which all panels are used once and landfilled, appears to be the least preferred option with respect to most impact categories. More detail on the results for these and other options can be found in references 4 and 5.

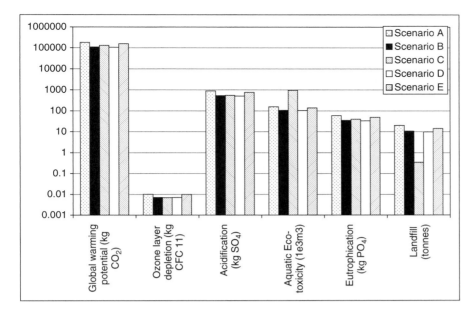

Figure 7.5 Comparison of life cycle impacts of different recycling scenarios for plastic panels

Key Facts 🐜——✂
● Closed-loop recycling, *i.e.* refurbishing and mechanical recycling, followed by landfill disposal of refurbished panels, is the most environmentally sustainable option for plastic panels assessed in this case study.

In summary, it appears that at present a combination of closed-loop recycling, *i.e.* refurbishment and mechanical recycling, followed by eventual landfilling of recycled panels is environmentally the most sustainable end-of-life option for plastic panels. These findings are based on the current technological, cost and market constraints that influence recycling of plastic materials. A change in these criteria would almost certainly change these conclusions and identify other end-of-life options as more sustainable.

7.3.2 Cascaded Use: Laminated Car Windscreens

The EC Directive on End-of-Life Vehicles (see Chapter 1) will lead to an increase in producer responsibility obligations for automotive material and component manufacturers. The Directive will encourage companies to increase their products' potential to be recycled or re-used, and consequently reduce escalating volumes of vehicle waste reaching landfills. Therefore, companies in this supply chain will need to be proactive with regard to the sustainable management of their materials and products. An example of what can be done in this respect is provided by the case study of laminated car windscreens, which is discussed next.

This case study compares different plastic materials that can be used for laminating car windscreens with the aim of identifying optimum end-of-life options for the laminate materials[5,6]. At the end of the useful life of a windscreen, only the glass is recycled leaving a growing amount of interlayer laminate waste to be disposed of. Currently, the polymer used almost exclusively in laminated-glass applications is poly(vinyl butyral) (PVB). Three other materials that could also be used as interlayers are also considered in this study: poly(vinyl chloride) (PVC), poly(ethylene-*co*-vinyl acetate) (EVA) and polyurethane (PU).

Figure 7.6 shows the life cycle stages in the production of laminated windscreens, from the production of the glass and polymer interlayers through to the windscreen shaping. The glass is produced in a float glass process and is transported to the laminating plant. After unpacking, the glass is cut to size. This operation produces approximately 30 % cullet (waste broken glass) which is transported back to the float glass production plant and stored for subsequent re-use in the float line. Once the glass has been cut to the correct size, the windscreens are made by pairing two sheets of glass together, with one sheet painted with

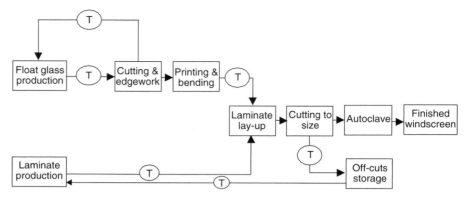

Figure 7.6 Life cycle diagram for the production of laminated windscreens (adapted with permission from Williams *et al.* (1999). Material and process selection methodology: a case study of laminated car windscreens. *7th SETAC LCA Case Studies Symposium.* Copyright (1999) SETAC-Europe, Brussels)

lead-free paint, before they are both bent to the appropriate angle. A temperature of 650 °C is necessary to obtain the required softening of the glass.

Each polymer is imported from abroad and incurs significant environmental impact, due to transportation. PVB requires additional energy input since it must be refrigerated to below 8 °C during both transportation and storage of the material. The interlayer is obtained in sheet form and the manual operation involves laying the polymer on top of one of the glass sheets in the pair, with the second sheet placed on top. Any excess laminate is simply cut away and recycled in the production of new laminate. The completed windscreens are then heated until the correct adhesion temperature is attained and the final product is inspected for faults. After the use phase, the glass is recycled and the interlayer disposed of in a landfill site.

For the purposes of comparison, the functional unit in this case study was defined as the production of 1000 finished windscreens. The environmental impacts of the first life cycle of the four polymer interlayers are compared in Figure 7.7. According to the ranking of the polymers shown in Table 7.1, the choice of the most environmentally sustainable material is not straightforward.

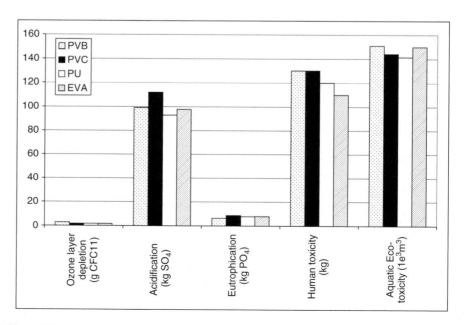

Figure 7.7 The first use of interlayer polymers: comparison of the life cycle environmental impacts

Table 7.1 Ranking of interlayers in order of preference regarding their life cycle environmental impacts (1, best option; 4, worst option).

	PVB	PVC	PU	EVA
Ozone layer depletion	2	1	1	1
Acidification	3	4	1	2
Eutrophication	1	3	2	2
Human toxicity	3	3	2	1
Aquatic toxicity	4	2	1	3

For example, PVC, PU and EVA all have the same ozone depletion figure, which is lower than that of PVB, so either of the three could be chosen as the best option for this impact category. PU is a preferred option for two more impacts: acidification and aquatic toxicity. EVA and PVB are, on the other hand, favoured with respect to human toxicity and eutrophication, respectively. However, overall, it could be concluded that the differences between the four materials are insignificant for the majority of environmental impacts. Hence, other criteria are going to play a more important role in identifying the most sustainable material for the first life cycle of these interlayers. For instance, economic analysis shows that PVC is the least expensive option for laminated windscreens. Technically, it delivers the same performance as the currently used PVB. Consequently, based on the assumptions used in this study, the use of PVC as a windscreen interlayer would appear to offer the best solution for the first life cycle.

The second part of this case study has considered the possibilities for further life cycles of the interlayer materials. The re-use of the interlayers in the same application is not practical because of the loss of their important performance characteristics (*e.g.* optical transparency) after the first life cycle. However, the polymers can be re-used in other applications and as an illustration, the study has evaluated the cascaded use of two candidate interlayer polymers: PVC and EVA. The aim was to find out whether the cascaded use alters the order of preference for polymers established after their first life cycle.

The cascaded option for PVC has involved the use of the PVC recyclate for pipe production while the option for EVA has considered its use as a cable jacket[5]. For each polymer two systems have been defined and their environmental impacts compared. Simplified flow diagrams illustrating these two systems are shown in Figure 7.8. The first system comprises the life cycle impacts of windscreen production and landfilling after the first use and the production of the pipes or cables from the respective virgin polymer. The second system consists of windscreen production and glass landfilling, with the PVC and EVA used to produce pipes and cables, respectively. The comparison is again based on 1000 windscreens with the assumption that 100 % of the polymeric material remaining after the first life cycle is recycled.

The environmental impacts of systems I and II are compared in Figure 7.9. It is obvious from the figure that for both materials the cascaded use results in lower environmental impacts than for the systems without re-use or recycling. The reason for this is that the cascaded systems are credited for the avoided burdens (and impacts) that would have been generated in the production of virgin polymers. The average reduction in the impacts for both materials is 20 %, with the highest difference found for aquatic toxicity. In the case of PVC, the cascaded use reduces this impact by 53 %, while for EVA this difference is even larger: 74 %. The results also show that the cascaded use of EVA yields the lowest environmental impacts, compared to the other three options. On average, the relative difference in the impacts between the cascaded EVA system and the next best option (cascaded PVC system) is 9 %. This leads to the conclusion that the use of EVA as a windscreen interlayer, followed by its cascaded use in cable jackets offers environmentally the most sustainable solution.

Key Facts
● Cascaded re-use of polymeric materials can reduce overall environmental burdens because it displaces the need for the virgin polymer and so eliminates the burdens from its production.
● Cascaded use can also be more economically viable than the use of virgin polymers.

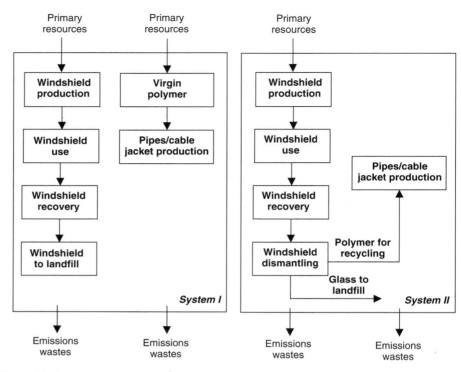

Figure 7.8 Cascaded use of interlayer materials: recycled PVC is used to produce pipes and EVA is used as a cable jacket

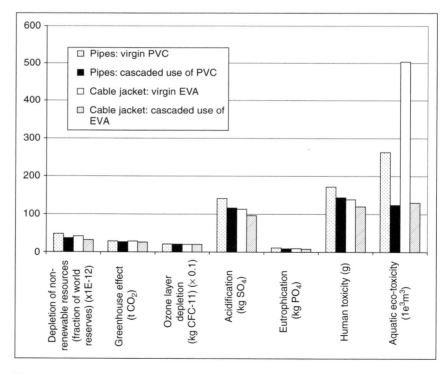

Figure 7.9 Comparison of life cycle impacts of the virgin and cascaded use of PVC and EVA

However, comparison of economic impacts for the cascaded use of these two materials indicates that PVC may be a preferred option. It offers an overall reduction in costs of energy and materials of 8 % compared to the system without material re-use. For EVA, this cost reduction is 5.3 %, resulting in a relative cost differential between the two systems of 34 %.

Thus, this study shows that both environmental and economic impacts can be reduced in a cascaded use of materials. Further reductions would be expected as the number of cascaded cycles increases. However, the study does not offer any definitive answers to the question of which option may be more sustainable overall. Nevertheless, it provides a transparent input into decision-making enabling decision-makers to trade off different sustainability criteria and to understand what exactly can be gained and lost with each option. In this particular case study, social concerns over the use of PVC related to emissions of dioxins from incineration (see Chapters 2 and 4), may influence the final decision on the most sustainable interlayer option.

7.3.3 Integrated Plastic Waste Management: Packaging

In addition to the ELV Directive, the EU Packaging and Packaging Waste Directive (see Chapter 1) also encourages more sustainable use of plastic materials. This Directive has provided an important driver for national governments and manufacturers to start devising integrated waste management policies, which would enable the achievement of these recycling targets. Consequently, a number of LCA studies have been initiated to assess the recovery and recycling options for waste packaging and to compare their environmental impacts. One of the more comprehensive studies was carried out for the European[1] and German[7] Plastics Industries and these results are discussed in this section.

This Germany-based study[1,7] has compared the environmental impacts of mechanical and chemical (feedstock) recycling and energy recovery from waste packaging. A simplified flow diagram of the recycling options considered in this study is given in Figure 7.10.

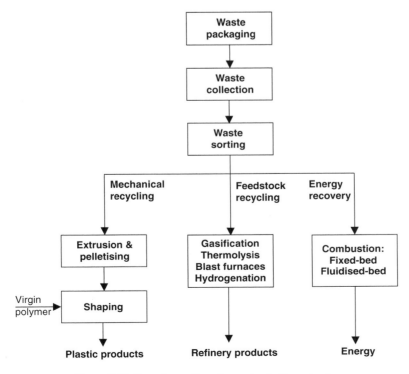

Figure 7.10 Recycling options for waste plastic packaging

Key Facts ⚷
● To compare
systems with vastly
different functional
outputs (*e.g.* a
plastic product and
energy), it is
necessary to expand
the boundaries so
that each system
includes both
outputs.

In 1995, approximately 525 000 tonnes of waste plastics were collected in Germany, of which 9 % was packaging film, 7.5 % containers, 58.4 % mixed plastics and 25.1 % sorting residue. Mechanical recycling has been considered for waste plastic bottles and film only and comprise the following options:

● recycling granulate from waste bottles back into the bottles;
● recycling packaging film back into the film;
● recycling film into waste sacks;
● recycling film into cable conduit.

Unlike mechanical recycling, feedstock recycling and energy recovery are more suitable for mixed plastics. The feedstock recycling technologies considered in this case study are:

● fixed-bed gasification with lignite;
● gasification with lignite in a fluidised bed;
● thermolysis of plastics into petrochemical products;
● use of plastics in blast furnaces;
● hydrogenation together with vacuum residue oils.

These options were described in more detail in Chapter 4; here we give a brief summary of these techniques, as a reminder. In the fixed-bed gasifiers, plastics, residues and lignite are partially oxidised at temperatures between 800 and 1300 °C and a pressure of 2.5 MPa. The main product is synthesis gas that can be used for example as a raw material for methanol synthesis. Fluidised-bed gasification is similar to that in a fixed-bed, except that the process is carried out in fluidised state at a higher pressure (3 MPa). The thermolysis plant converts packaging material into petrochemical products. If waste plastics are used in the blast-furnace process, they replace some of the heavy oil used to generate CO and H_2, which reduce iron ore to iron. The remaining plastics can also be used for heat generation in this application. Finally, hydrogenation converts waste plastics with hydrogen into a synthetic crude oil, which is further processed in the petrochemical industry.

Energy recovery options have included combustion in fixed and fluidised beds. In both cases, 26 MJ of energy is generated *per* kilogramme of recovered plastics. Of that amount, 17 % is converted to electricity, the same percentage is recovered as useful heat and the remaining 66 % is removed with cooling water.

The environmental impacts of different recycling options are compared in two stages. The first stage examines the feedstock recycling and energy recovery options and the second stage compares these methods with mechanical recycling. However, each of the recycling options delivers a different service or produces a different 'product', which makes direct comparison difficult. For, example, in mechanical recycling, 1 kg of plastic material is processed into approximately 1 kg of product (*e.g.* bottles or plastic film). The 'product' from energy recovery, however, is 26 MJ of heat for the same amount of input waste.

It is, nevertheless, possible to compare these systems, but only after some modification. Each system must first be expanded to include an equivalent process generating the same 'product' as the system with which it is being compared, so that they deliver the same functional units. For example, as illustrated in Figure 7.11, if the production of mechanically recycled plastic film is to be compared with energy recovery, then the recycling system is expanded to include an alternative process for heat generation. In that way this system produces two functional outputs: packaging film and heat. Similarly, the boundaries of the energy recovery system are expanded to include an equivalent process for film manufacture so that it now delivers the same functions as the mechanical recycling system. System expansion carried out in this way is equivalent to avoiding allocation in LCA (discussed in Chapter 6).

The results of the first stage in which feedstock recycling and energy recovery are compared are shown in Figure 7.12. Landfilling has been chosen as the reference scenario, so that the results show a difference between the recycling options and landfilling. For example, the use of plastics in blast furnaces saves 29.3 MJ energy *per* kilogramme of waste

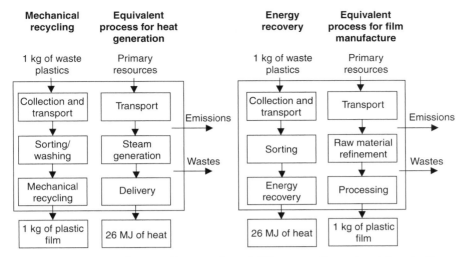

Figure 7.11 System expansion to enable comparison of different recycling options (adapted with permission from Brandrup (1998), 'Ecological and economical aspects of polymer recycling.' In *Macromol. Symp.*, **135**, pp. 223–235. Copyright (1998) Wiley-VCH, Weinheim

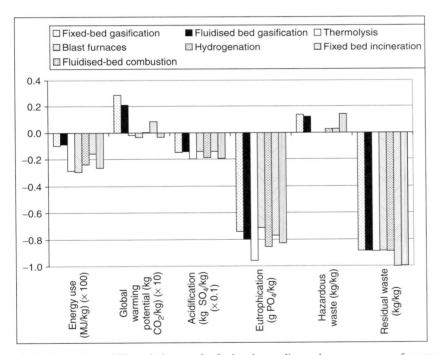

Figure 7.12 Comparison of life cycle impacts for feedstock recycling and energy recovery from waste plastic packaging

packaging compared to energy used when the plastic waste is landfilled. The options are ranked according to their environmental impacts in Table 7.2.

As you can see from Figure 7.12, all feedstock and energy recovery options have lower environmental impacts than landfilling. The exception to this are two impacts: global warming, which is increased by gasification, hydrogenation and fixed-bed incineration, and hazardous waste, which is higher than for landfilling for all but two options (thermolysis and blast furnace). Regarding energy consumption and global warming potential, the use of waste plastics in blast furnaces appears to be the best option (see Table 7.2), followed closely by thermolysis and fluidised-bed combustion. However, with respect

Table 7.2 Feedstock recycling and energy recovery options ranked with respect to their environmental impacts (1, best option; 7, worst option).

	Energy	Global warming	Acidification	Eutrophication	Hazardous waste	Residual waste
Fixed-bed gasification	6	6	4	6	6	6
Fluidised-bed gasification	7	7	5	4	5	5
Thermolysis	2	2	1	1	1	3
Blast furnaces	1	1	6	7	2	6/7
Hydrogenation	4	4	3	2	3	4
Fixed-bed incineration	5	5	7	5	4	1
Fluidised-bed combustion	3	3	2	3	7	2

to acidification, eutrophication and hazardous waste, thermolysis is environmentally the most sustainable option. Hydrogenation also compares favourably for these three impacts, but is less favourable for energy use, global warming and residual waste. Fluidised-bed combustion, which scores highly for most of the other impacts, is the least preferred option regarding hazardous waste. The reason for this is the greater mass of filter dust generated in the combustion process, which has to be disposed of. In summary, under the conditions considered in this case study, the feedstock recovery of plastics in blast furnaces and thermolysis could be recommended as the most sustainable options.

In the second stage of this study, mechanical recycling of waste bottles and film has been compared with feedstock recycling and energy recovery, again using landfilling as the reference scenario. These results are shown in Figure 7.13. Firstly, it can be noticed that there is an overall reduction in the impacts for all mechanical recycling options compared to the reference scenario. Furthermore, the best option for all impacts appears to be film recycling into refuse sacks.

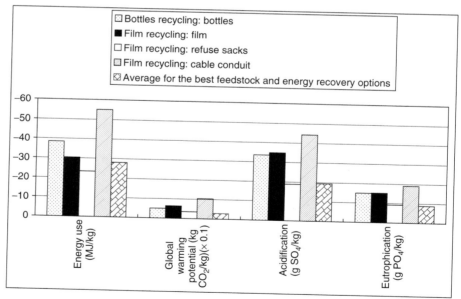

Figure 7.13 Comparison of life cycle impacts from recycling of bottles and film with the best options for feedstock recycling and energy recovery

A comparison of mechanical recycling with the average values for the best feedstock and energy recovery options is also shown in Figure 7.13. Overall, mechanical recycling results in lower environmental impacts than either the feedstock or energy recovery. The only exception to this is the amount of residual and hazardous waste from the four mechanical recycling options (not shown in the figure). These values range from 160 to 220 g kg^{-1} of recovered plastics, which is higher than for any feedstock or energy recovery option. Further details on this case study can be found in APME[1] and Brandrup[7].

In summary, the findings of this study show that mechanical recycling is environmentally more sustainable than either feedstock or energy recovery. However, given the capacity, technological and sorting constraints at present, it is more likely that integrated waste management, combining mechanical and feedstock recycling with energy recovery of packaging that cannot be recycled is going to be the most practicable environmental option.

7.3.4 Life Cycle Product Design for Chemical Recycling: 'Waterlily' Cushioning

The previous case studies have concentrated on the end-of-life options for different plastic products. Following on from the discussion that we presented in Chapter 6, the case study discussed in this section further demonstrates the advantages of the design for the environment, where the focus is shifted from the end to the beginning of the life cycle of a product. This case study, carried out by Markovic and Hicks[8], applies the life cycle design principles to develop a novel, recyclable polyurethane (PU) furniture cushioning material (mattress) called 'Waterlily'. The aim of the 'Waterlily' project has been to develop a product with reduced environmental impacts along the whole life cycle. The study has been divided into two stages. The first has involved identification of the main stages in the life cycle to be targeted for maximum improvements with respect to environmental performance. The second stage has aimed to identify the most appropriate end-of-life options for PU foam, which would enable redesign of the existing product for improved recyclability.

The LCA results for the energy consumption of the existing product are summarised in Table 7.3. They show that nearly 80 % of the total energy is consumed upstream, *i.e.* from extraction of oil from the ground up to the manufacture of the main raw materials. The raw materials used for the majority of PU foams include tolylene-2,4-diisocyanate (TDI) and a polyol which, in the presence of water, evolve carbon dioxide (the blowing agent) which expands PU into a foam.

Table 7.3 Summary of LCA results for flexible PU foam mattresses[8]. Reproduced with permission from Markovic and Hicks (1997). 'Design for chemical recycling.' In *Philos. Trans. R. Soc. London Ser. A*, **335**, pp. 1415–1424. Copyright (1997) Royal Society.

Life cycle stage	Energy use (%)	Key environmental issues
Raw material manufacture	80	Global warming
		Acidification
		Water pollution
		Toxic chemicals
Product fabrication	5	(Workers' safety)
		Ozone depletion
		Solid waste
Packaging and transport	5	Global warming
		Acidification
Use	0	(Consumer safety)
Waste management	10	Recyclability
		Volume and toxicity of waste

However, the analysis of other environmental impacts did not point out clearly which life cycle stage carried the greatest burden. Instead, each stage was associated with different environmental problems, which made it more difficult to focus the efforts on the most important life cycle stages in redesigning the existing product. Nevertheless, the main benefit of this exercise has been to draw attention to the importance of the manufacture of raw materials and hence the effort has subsequently focused on redesign for recycling.

In the second stage of the project, key design criteria for each life cycle stage have been identified; these are summarised in Table 7.4.

Key Facts
● The major environmental impacts of a plastic product are generally associated with extraction of raw materials and synthesis of the polymer.

Table 7.4 Life-cycle design criteria for 'Waterlily' mattress[8]. Reproduced with permission from Markovic and Hicks (1997). 'Design for chemical recycling.' In *Philos. Trans. R. Soc. London Ser. A*, **335**, pp. 1415–1424. Copyright (1997) Royal Society.

Life cycle stage	Key design parameters
Raw material manufacture	Less volatile raw materials
	Simpler formulation
Product fabrication	No isocyanate vapour
	No organic blowing agents
	No autocombustion
Packaging and transport	Recyclable
	Minimum impacts from transport
Use	Improved fire performance
	No halogens
	Comfort performance
Waste management	Designed for recycling (mechanical, chemical and energy recovery)

For the raw material manufacturing stage, the main aims have been to develop a simpler formulation and to replace tolylene-2,4-diisocyanate (TDI), of relatively high volatility, with lower-volatility 4,4'-methylene-*bis*(phenyl isocyanate) (MDI) and so improve working-environment conditions. In the mattress production stage the main requirements were to minimise isocyanate vapour in the workplace, avoid use of organic (ozone depleting) blowing agents and eliminate the possibility of autocombustion. Furthermore, the packaging material should be recyclable and the environmental impacts of transportation of low-density foam should be minimised. Regarding the mattress use phase, the objective has been to reduce fire risk without the use of halogenated fire retardants and to achieve superior comfort performance. Finally, the waste management stage should enable mechanical and chemical recycling or energy recovery as appropriate.

Several recycling options have been considered. At the time this study was carried out, much of the flexible foam was exported to the USA for rebonding scrap chips into carpet underlay. However, there was a concern that the export of foam scrap into the USA could exceed the capacity of the rebond market and that the prices would continue to decrease. The second recycling option examined has been mechanical recycling to a fine powder, which can be added as a filler to the polyol component, used to produce flexible foam. The problem associated with this option is that the incorporation of filler often affects the physical properties of the foam so that only 10–20 % can be added. The next possibility has been energy recovery by incineration, but addition of halogenated species may contribute to the formation of dioxins during the combustion process. The final option considered in this case study has been chemical recycling which breaks down the PU into basic chemicals, monomers or hydrocarbon feedstock. The concern here has been that the presence of nitrogen in PU reduces the value of the monomer in feedstock recycling. However, depolymerised PU can still be re-used in the same application. Hence, a preliminary decision has been made to develop a new 'Waterlily' mattress which would incorporate all of the above environmental improvements and would be designed for closed-loop chemical recycling by a process known as split-phase glycolysis (a process similar to alcoholysis, which will be further discussed in Chapter 8). In this process, compacted pellets of the used 'Waterlily' mattress are dissolved with diethylene glycol (DEG) in the presence of a catalyst. The reaction mixture is then allowed to separate into two layers. The bottom layer consists of DEG and aromatic compounds and can be used to make rigid foams. The top layer, which consists of DEG and flexible polyol, is then purified with more DEG to give pure flexible polyol that is used to replace virgin polyol completely in the production of new 'Waterlily' mattresses.

To check that the proposed chemical recycling option is indeed sustainable, the four recycling methods have been compared from 'cradle to grave', including the manufacture

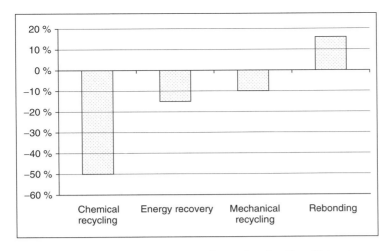

Figure 7.14 Comparison of different recycling options for 'Waterlily' mattresses

of the raw materials and production of mattresses, the collection of the used mattress in a segregated collection scheme and the recycling options themselves. The results, showing energy use, are summarised in Figure 7.14 and demonstrate that chemical recycling can save up to 50 % of the energy required to manufacture the mattress from virgin materials. Energy recovery and mechanical recycling would result in a saving of up to 15 % and 10 %, respectively, compared to the original energy used. On the other hand, rebonding foam chips for carpet underlay consumes approximately 16 % more energy than manufacturing of the virgin mattresses.

Therefore, design for chemical recycling in this case appears to be the most sustainable option. However, this option is not without its challenges. One of these is setting up and maintaining a recycling loop involving collection, sorting, cleaning, transport and disassembly. This requires commitment of all parties involved, from the producers of PU foam through furniture manufacturers to retailers, consumers and policy makers. These challenges can, however, be resolved by setting up new alliances in the supply chain in which there is a concerted effort of all the parties involved and in which each group of stakeholders can benefit. Without that, products such as 'Waterlily' will remain the isolated examples of what could be done to improve sustainability of polymers, with little prospect for practical application.

7.4 SUMMARY AND LEARNING OUTCOMES

In this chapter we have illustrated, with selected case studies, how different recycling options impact on the environment. The presented results show that the impacts of recycling are very much dependent on the particular situation and that it would not be appropriate to draw generic conclusions without considering the specific conditions on a case-by-case basis. The results also show that in most cases one option will be better for some environmental impacts but worse for the others and that trading off between the impacts, economic costs and other criteria will be necessary to make the most appropriate decision for a particular situation.

Upon studying the material presented in this chapter, you should be able to understand and discuss:

- the life cycle implications and environmental impacts of recycling;
- the advantages and disadvantages of each recycling option with respect to their general environmental impacts;
- environmental impacts of recycling options for specific case studies presented in this chapter;

● how technical, environmental, economic and social criteria influence the decision-making process in choosing the most sustainable recycling option.

7.5 REFERENCES AND FURTHER READING

1. APME (1995). *Life Cycle Analysis of Recycling and Recovery of Households Plastics Waste Packaging Materials. Summary Report*, Association of Plastics Manufacturers in Europe, Brussels.
2. BUWAL (1991). *Ecobalance of Packaging Materials, State of 1990*, Federal Office of Environment, Forest and Landscape (BUWAL), Switzerland.
3. Hunt, R.G. (1995). LCA considerations of solid waste management alternatives for paper and plastics, *Resour. Conserv. Recycling*, **14**, 225–231.
4. Freire, F., Williams, E., Azapagic, A., Clift, R., Mellor, W. and Stevens, G.C. (2002). Life cycle activity analysis: a case study of plastic panels. Part III: environmental concerns. In: Technology Commercialisation: DEA and Related Analytical Methods for Evaluating the Use and Implementation of Technological Innovation (Thore, S.A.), Kluwer, Dordrecht, pp. 232–352
5. Mellor, W., Williams, E.A., Stevens, G.C., Clift, R. and Azapagic, A. (2001). *Chain Management of Polymer Materials (CHAMP) Final Report*, University of Surrey, UK.
6. Williams, E.A., Mellor, W., Azapagic, A. Stevens, G.C. and Clift, R. (1999). Material and process selection methodology: a case study of laminated car windscreens. 7th SETAC LCA Case Studies Symposium, SETAC-Europe, 2 December, Brussels.
7. Brandrup, J. (1998). Ecological and economical aspects of polymer recycling, *Macromol. Symp.*, **135**, 223–235.
8. Markovic, V. and Hicks, D.A. (1997). Design for chemical recycling. *Philos. Trans. R. Soc. London Ser. A*, **355**, 1415–1424.

7.6 REVISION EXERCISES

1. Based on the examples presented in Chapter 6, which stages in the life cycle of a product do you think impact most heavily on the environment? Why do you think that this is so?

2. Based on the examples presented in this chapter, which life cycle stages in recycling impact most heavily on the environment? Why do you think that this is so?

3. Which environmental impacts can you expect to arise from polymer recycling? Discuss each recycling option separately and support your discussion with specific examples.

4. In the case study of plastic panels discussed in Section 7.3.1, incineration with energy recovery was not considered. How do you think the conclusions of the study would change with respect to the ranking of different options if incineration *without* energy recovery were to be replaced by incineration *with* energy recovery?

5. The laminated windscreens case study did not offer any definitive answers as to which polymer may be a better option for the interlayers. Considering the technical, economic, environmental and social factors that influence the choice in this case, discuss which polymer you would choose and give reasons to support your choice.

6. Why is integrated waste management the best practicable option for plastic packaging? Support your answer by taking into account all three components of sustainable development, *i.e.* environmental impacts, economic costs and social factors.

7. In the case study on the 'Waterlily' mattress, why did the company decide to concentrate on its design for chemical recycling? What were the drivers that compelled the company to consider the life cycle impacts of this product? What are the barriers against the practical application of this option?

8. How could the Life Cycle Product Design methodology presented in Chapter 6 be used in the design of the 'Waterlily' mattresses?

Chapter 8 – The Damsel of the Sanct Grael, (DG Rossetti, 1874)

The grail myth of Arthurian legend is an enduring one and has been depicted on many occasions, rarely so strongly. The grail quest represents the search for the ultimate goal of enlightenment – in this case the development of an approach that enables society to take the necessary steps towards a sustainable future. It would have been easy to select another image by Waterhouse or Burne-Jones, as both painted similar subjects, but unthinkable not to have included another painting by Rossetti in the book.

future directions: towards sustainable technology

8.1 INTRODUCTION

Throughout this book we have laid out the arguments for the need to become more sustainable in the way we use resources and manage polymer waste. We have discussed the issues that drive and limit sustainable practices and described a number of technical, economic, environmental and social factors that need to be addressed in order to achieve this aim. Clearly, we have a long way to go before we can state that sustainable resource and waste management is the norm. In this chapter we explore further how to respond to this challenge and reduce the impact of polymers on the environment. The development of new sustainable technologies and processes is certainly one of the ways to move forward and we first examine a variety of methods that are currently being considered, both for the synthesis of new and for recycling of used polymers. However, a solution that is wholly based on technology is unlikely to be a panacea and we will have to look at this problem in a much broader way. This includes the development of appropriate national and international policies and also rethinking of our consumption patterns and changing our life styles. These and other non-technological issues have been encountered throughout the book and are briefly reiterated at the end of this chapter.

8.2 IMPROVING THE EFFICIENCY OF POLYMERISATION: THE PRINCIPLES OF GREEN CHEMISTRY

Green chemistry is an emerging concept aimed at identifying processes and pathways that enable more sustainable synthesis of chemicals. Anastas and Warner[1] have formulated twelve principles of green chemistry, encompassing all stages of the design of a new chemical product (see Textbox 8.1). For example:

Textbox 8.1 Anastas and Warner's twelve principles of green chemistry[1]

1. Waste	It is better to prevent waste than to treat or clean up waste after it has been formed.
2. Synthesis	Synthetic methods should be designed to maximise the incorporation of all materials used in the process into the final product.
3. Toxicity	Wherever practicable, synthetic methodologies should be designed to use and generate substances that possess little or no toxicity to human health and the environment.
4. Efficacy	Chemical products should be designed to preserve efficacy of function while reducing toxicity.
5. Auxiliaries	The use of auxiliary substances (*e.g.* solvents, separation agents, *etc.*) should be made unnecessary wherever possible and, innocuous when used.
6. Energy	Energy requirements should be recognised for their environmental and economic impacts and should be minimised. Synthetic methods should be conducted at ambient temperature and pressure.
7. Feedstock	A raw material of feedstock should be renewable rather than depleting wherever technically and economically practicable.
8. Derivatisation	Unnecessary derivatisation (blocking group, protection and deprotection, temporary modification of physical/chemical processes) should be avoided whenever possible.

Key Facts
- It is unlikely that technological improvements alone will resolve the issue of sustainability and we shall have to look in the future to our lifestyles and patterns of usage.
- Sustainable methods for chemical synthesis minimise use of energy and production of waste and maximise efficiency and yield.

9. Catalysis	Catalytic reagents (as selective as possible) are superior to stoichiometric reagents and contribute to increasing the reaction efficiencies.
10. Degradation	Chemical products should be designed so that at the end of their function they do not persist in the environment and break down into innocuous degradation products.
11. Monitoring	Analytical methodologies need to be further developed to allow for real-time, in-process monitoring and control prior to the formation of hazardous substances.
12. Accidents	Substances and the form of a substance used in a chemical process should be chosen so as to minimise the potential for chemical accidents, including releases, explosions and fires.
13. Simulation*	Where possible, the reliable simulation or prediction of materials properties should be employed to optimise synthesis and reduce impact on the environment.

*Added by the authors of this book.

Key Facts
● Molecular simulation allows the investigation of processes and optimisation of conditions before experimentation starts for real.

- use of toxic materials and waste generation should be minimised;
- energy and raw materials should be renewable and waste materials easily degradable;
- benign catalysts should be used to promote reaction efficiency;
- the use of auxiliary materials (*e.g.* solvents) should be reduced.

To this 'clean dozen' we should also like to add a thirteenth principle: molecular simulation, which aims to optimise synthesis and hence reduce the impact on the environment. Although some of these principles, which we have already discussed in several places in the book, may be self-evident, common sense is rarely common practice and this approach heralds the way to the necessary future of materials synthesis. We will now follow some of these principles and discuss how they can be applied for cleaner monomer synthesis. We start with simulation, which may be viewed as the first step to a clean synthesis.

8.2.1 Molecular Simulation

The development of new monomers and modified macromolecules has stimulated much interest in both academia and industry and a new generation of, predominantly organic, chemists are applying their considerable expertise in synthesis towards this endeavour. In common with 'traditional' chemical synthesis involving the preparation of 'small' organic molecules, polymer synthesis brings many similar challenges and a few additional problems (*e.g.* limited solubility or reduced reactivity) arising from the extended chain structure. However, no matter what the nature of the preparation, every synthesis will require that reaction conditions (*e.g.* temperature, solvent, pressure, reagents, and catalysts, *etc.*) are optimised. Furthermore, every preparation will consume energy and feedstock materials and generate waste (in the form of gaseous emissions, solvated (by)products or solid matter; the latter may be the desired product, but might also be accompanied by contaminants). While still in its infancy in the field of polymer chemistry, computational chemistry has the potential to simulate the properties of novel materials, as yet unsynthesised. Ultimately, this would allow the scientist or engineer to examine the potential of a novel structure to give him or her the required properties (*e.g.* heat resistance, bulk modulus or glass transition temperature) to fulfil a new application, without having to prepare the actual material in the laboratory. Although computation experiments do consume energy, the additional burdens that synthesis places on the environment would be largely avoided or dramatically reduced. At a later stage, the new monomer or polymer would be prepared in order to allow small scale testing, prior to scale up, but extensive exploratory syntheses to develop many potential monomers would be unnecessary.

The increasing advances made in computational power make the use of simulation techniques increasingly useful in the rational design of new materials. The subject is growing rapidly and a thorough discussion is beyond the scope of this book, but you can read more about this interesting topic and its application to chemistry in Goodman's book[2]. Since its first application to the pharmaceutical industry to model the interaction of drug molecules with the active sites of proteins, molecular modelling has become generally accepted as a powerful technique for understanding some of the properties exhibited by materials. There is an increasing awareness and acceptance of the place of computing in chemistry (and indeed in daily life). Consequently, the use of general modelling techniques (*e.g.* semi-empirical molecular mechanics and molecular dynamics methods) is more widely accepted and is beginning to be applied more frequently to yield reliable simulations of physical, mechanical, and electrical properties of polymers. There are, however, still a couple of fundamental problems to hinder the wider use of molecular modelling in this area.

Key Facts
- The computational chemist considers a polymer as a flexible scaffold of balls (atoms) connected by springs (bonds).
- The configuration that minimises steric hindrance and bond strain is regarded as the stable structure of the system.
- Modelling of polymers is computationally very demanding, because of the large number of atoms *per* molecule.

As noted in Chapter 2, most synthetic polymers (although not all) tend to be either amorphous or semi-crystalline. This can present a significant problem when modelling polymers, as the parameters used for molecular modelling for bond lengths, bond angles and torsion angles are normally taken from crystal structures and must be representative of the structure under consideration. The computational chemist attempts to construct a 'balls and springs' (molecular mechanics method) model by considering the polymer as a flexible scaffold comprising spherical atoms. They are joined to each other through specific joints by bonds of various lengths depending on whether they are 'single' bonds (*e.g.* $-O-H$) or 'multiple' bonds (*e.g.* $-C\equiv C-$). The joints also contain different numbers of junctions to represent tetrahedral sp^3 (*ca.* 109.47°), trigonal sp^2 (*ca.* 120°), or planar sp (180°) hybrid orbitals. The balls are distinct from each other in their *atom type* (*i.e.* not all the atomic species of the same element are equal, unlike traditional quantum mechanics methods). In practice, this means that the mathematical functions used in the energy calculation contain different parameters for the same element depending on its surroundings, *e.g.* hybridisation, valence bonding, neighbouring atoms, *etc.*

The molecular mechanics method assumes that the forces experienced by an atom can be calculated from a potential energy surface: the force field. This force field describes the total energy due to bond stretching, bond angle bending, torsional rotations and nonbonded interactions (such as Coulombic forces and van der Waals' interactions). Once the molecular framework has been constructed, the 'balls and springs' are caused to flex and rotate to enable the total energy of the system to reach the lowest possible state. This 'global energy minimum' represents the most stable chemical structure with *e.g.* steric hindrance and bond strain at a minimum. This is not always as simple as it sounds as often local minima can be obtained during the calculation which may lie very close to the unique global minimum and a variety of minimisation methods may be employed to achieve convergence to an energy minimum. Figure 8.1 shows the effect of rotating a torsion angle (ϕ_3) on the energy of the ether bridge (shown in the detail).

The second major stumbling block to the wider application of modelling to polymer chemistry has been due to the fact that the chain lengths of polymers are generally very large and are composed of many thousands or even millions of atoms. For instance, during a molecular dynamics simulation, Newton's equations of motion are used to follow the movement of the model, *i.e.* the trajectory or molecular motions, as a function of time. Molecular dynamics can be used to study the behaviour (*e.g.* diffusion or folding) of a polymer by raising the temperature by supplying kinetic energy to the model. New velocities and positions of each of the atoms are then calculated for each time step (again using Newton's laws of motion). As the simulation is basically a series of mathematical calculations in which the trajectory of every atom is being calculated for every time step, a large molecule (*i.e.* a macromolecule) is computationally very demanding, requiring a powerful computer unless calculations are to be prohibitively slow.

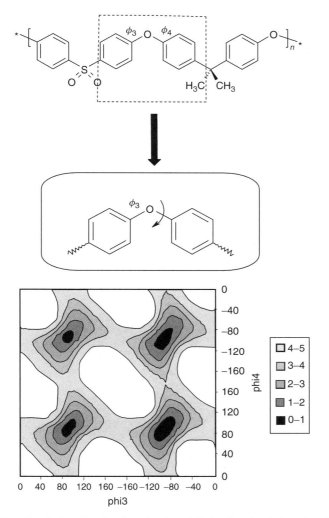

Key Facts
● Computed simulations generate mathematical models that must be tested against the properties of known polymers.

Figure 8.1 Conformational plot of torsional angles ϕ_3 and ϕ_4 for the ether linkage in a thermoplastic poly(arylene ether sulphone) showing energy minima (the data are contoured in height in units of kcal mol^{-1}) (data originally published in reference 3)

Fortunately, there have been significant increases in computational power of late, making supercomputers (or now even 'metacomputers' with a number of PCs connected into a network) available to many laboratories and increasingly sophisticated personal computers within the reach of many individuals.

Consequently, the use of general modelling techniques is more accepted and these are beginning to be applied more frequently to yield reliable simulations of physical, mechanical and electrical properties of polymers. To date, a range of thermoplastics and even thermosetting polymers have been studied (Figure 8.2)[3-5]. The latter are more complex to study as the production of a representative structural repeat unit is somewhat more difficult to achieve.

However, it must be recognised that the simulation model is precisely that, *a mathematical model*. In order to test its reliability in simulating the empirical properties of interest, it should be validated against the properties of known polymers (and the greater the 'training set' the higher the confidence in the model) before any prediction may be attempted. The potential benefits for a reliable simulation or prediction of materials properties are obvious. By simulating the properties of as yet unmade polymers it should be possible to screen materials for beneficial characteristics and thus optimise synthesis. In this way some of the

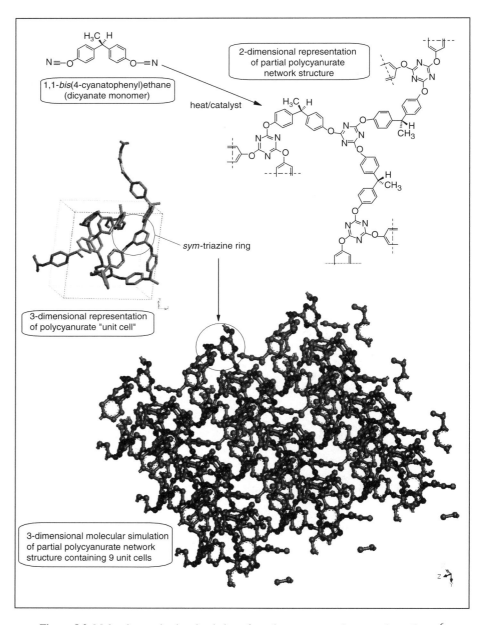

Figure 8.2 Molecular mechanics simulation of a polycyanurate, a thermosetting polymer[6]

chemical impact on the environment, *e.g.* the disposal of waste solvents and raw materials usage, may be significantly reduced.

8.2.2 Alternative Feedstocks

As outlined in Chapter 2, the polymer industry has historically relied heavily on nonrenewable, fossil fuel resources for the necessary feedstock materials. Almost all polymers in current use, such as plastics, rubbers and fibres are synthesised from chemicals derived from oil (see Figure 2.2 in Chapter 2). The chief exceptions to this rule are natural rubber and related polymers, and polymers derived from cellulose, such as cellulose acetate. In 2000/2001, the world production of oil was around 65.5 million barrels *per* day or

2.4×10^{10} barrels *per annum*. As noted in Chapter 1, around 4% of the contents of that are used in the production of plastics, which means that 960 million barrels of oil *per annum* are used in the production of plastics alone. Coupled with the energy and material inefficiencies, which accompany the use of fossil fuels, our continuing reliance on nonrenewable resources clearly represents an unsustainable practice and alternative feedstock materials must be found.

For instance, there is great potential in agricultural feedstocks such as Kraft black liquor, which can yield peracetic acid (a useful precursor to a number of monomers including some epoxy resins) *via* hydroxy acids and acetaldehyde, or anthraquinone *via* lignin). Furthermore, potato waste can yield acrylic acid or peracid esters (which can also be transformed into monomers) *via* the fermentation of sugars. Alternatively, biological feedstocks may be of even greater interest. The treatment of biomass by pyrolysis, gasification or fractionation techniques yields some 25 useful compounds, including butadiene, tetrahydrofuran, and toluene. Gandini and Belgacem[7] have reviewed the production of polymers derived from biomass components for a wide range of technological applications. Similarly, Donnelly[8] recently reviewed the enormous progress in the *in vitro* use of enzymes for the synthesis of polymers containing saccharides, lignins, proteins and related compounds. He has outlined the wide range of, often-biodegradable, polymers produced of both academic and technological interest. The particular benefits of the use of enzymes that have been highlighted include the potential for a high degree of control over the polymer architecture produced. Donnelly also raised the issue of environmental impact associated with the *in vitro* use of enzymes and stated that consideration needs to be given to a variety of questions, such as:

- Does modification of the enzyme allow retention of its biodegradability?
- Are the auxiliary materials used (or reaction byproducts) biodegradable, or do they need to be recovered and recycled efficiently to minimise environmental impact?
- What is the likely environmental impact of the use of auxiliary materials or reaction byproducts?
- Can the polymeric product be tested (or has it been tested) in a meaningful way to assess its environmental compatibility?

Some of these methods are at a very early stage of development. They may be uneconomic at present, but such is the need for change that the degree of political pressure in the form of legislation or taxation/fiscal benefits will increase. This will certainly encourage further development in this area, which may eventually lead to our reduced reliance on fossil fuels.

8.2.3 Auxiliaries: Reducing our Reliance on Organic Solvents

Supercritical Fluids (SCFs)

A number of alternative strategies may be adopted for synthesis of materials, involving the development of alternative reagents or solvents or, alternatively, the development of solvent-free processes. For instance, we will talk about SCFs again in the next section in the context of waste management, but SCFs are also attracting increasing interest as environmentally friendly alternatives to conventional organic solvents. SCFs may display the diffusivity of a gas (an important feature when considering reaction kinetics) while having the density of a liquid (facilitating the solvation of many compounds). Of the common materials outlined in Table 8.1, supercritical carbon dioxide (scCO$_2$) is readily accessible under relatively mild conditions (T_c of 31 °C and a P_c of 73 atm, see Figure 8.3) and is also abundant (*e.g.* from fermentation), inexpensive, nonflammable and nontoxic. Interest in the use of SCFs in polymer synthesis is growing and scCO$_2$ has been used as the continuous phase for all of the main types of chain growth and step growth polymerisation mechanisms, including metal-catalysed, free radical and ionic

> **Key Facts**
> - The *in vitro* use of enzymes to synthesise polymers has the potential to reduce the use of fossil fuels.
> - Supercritical CO$_2$ is cheap and abundant and could replace chlorinated solvents for synthetic processes.

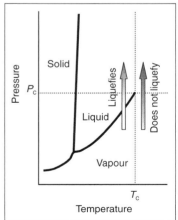

Figure 8.3 A phase diagram showing the physical properties of SCFs (redrawn from Atkins, P. W. and Jones, L. (1999). *Chemical Principles. The Quest for Insight*, Freeman, New York, p. 357. Copyright (1991) Freeman Publishers)

Table 8.1 Conditions for critical behaviour for selected common substances.

Substance	Critical pressure (atm)	Critical temperature ($^{\circ}$C)
H_2	13	−240
O_2	50	−118
H_2O	218	374
N_2	34	−147
CO_2	73	31
CH_4	46	−83

Data adapted from Atkins, P. W. and Jones, L. (1999). *Chemical Principles. The Quest for Insight*, Freeman, New York, p. 357. Copyright (1999) Freeman Publishers.

Key Facts
● Between 2 and 7 % of solvent (toluene) is released to the environment during printing.
● Aqueous-based polymer coating compositions are often inferior to organic-based equivalents, but can be improved by the use of cross-linking agents.

processes[9]. Polymers also tend to become plasticised in $scCO_2$ so that the T_g is significantly lowered, thus making the removal of residual monomer from the polymer (or introduction of additives or formation of foams) much easier to achieve. However, care has to be taken in its use because of its global warming potential. Assessed on a life cycle basis, the use of supercritical CO_2 may exceed its benefits (because of the high degree of energy usage) so that LCA should be carried out to ensure that overall, the use of this fluid is sustainable.

Water-mediated Polymerisation

Currently, solvent-borne inks are preferred for publication printing, using the Gravure process, usually involving organic solvents such as toluene. However, the concerns over the emissions of volatile organic compounds (VOCs) during the use of solvents and their global warming and ozone depleting potentials have led to a need to replace the organic solvents with water-based solvents. It has been estimated that between 2 and 7 % of the solvents (*e.g.* toluene) contained in inks is released during the printing process[10]. Currently, there are two commonly used waterborne polymeric systems for coating applications: poly(vinyl alcohol) and polymethacrylates (the latter may contain both acrylate and methacrylate monomers). Acrylic systems possess superior qualities in terms of ease of processing, endurance of modification, and generally excellent film forming properties. Their use is becoming increasingly widespread in a variety of coating compositions including paints, varnishes, printing inks, adhesives and textile finishing formulations.

In an emulsion polymerisation system, monomer(s) may be distributed either as stabilised droplets, solubilised in surfactant micelles, or dissolved to some extent in the aqueous phase where initiation can take place. Polymerisation is thought to take place within these micelles (Figure 8.4), with monomer droplets serving to replenish the supply of monomer to the polymerisation sites by diffusion through the aqueous phase. As the micelles grow, they adsorb free surfactant, generally containing a polar functional 'head' group and a nonpolar, hydrophobic 'tail') from the solution and, eventually, from the surface of the emulsion droplets. Through this process, the polymer particles in the final latex are stabilised by the surfactant. The relative concentrations of the surfactant and monomer(s) can be varied to alter the number and size of polymer particles ultimately produced and the rate of polymerisation.

Aqueous-based (oil-in-water) polymer coating compositions form films through coalescence as the aqueous continuous phase evaporates. A homogeneous, clear and glossy film is obtained when coalescence proceeds smoothly, although the addition of a small amount of a coalescing agent may be required to act as a lubricant for more rigid polymer chains. Aqueous-based polymer coating compositions, which rely on particle coalescence for film formation and strength, can be inferior to more traditional solvent-based nitrocellulose systems, solvent resistance and film strength being their main weaknesses. However, commercial coating compositions have incorporated functional groups (*e.g.* carbonyl groups) along a copolymer backbone, which may undergo cross-linking reactions with polyhydrazides[11,12] or polyhydrazines[13] *via* the elimination of water. A similar method

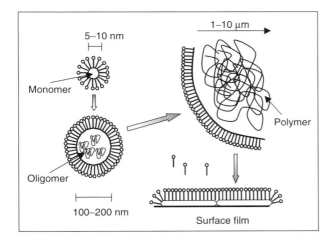

Figure 8.4 Schematic to show emulsion polymerisation *via* micelle formation and coalescence

Key Facts

● Ziegler–Natta catalysts have been used for 40 years to control tacticity of polymers, but now metallocenes allow chirality to be controlled also and a wide range of properties can be achieved from a single polymer.

involves the incorporation of hydrazide groups in the chain to undergo reaction with di- or poly-ketones. The resulting cross-linked polymers yield water-resistant films, although recent research has been directed at removing potentially harmful hydrazines from the formulations while enabling the films to form cross-links at ambient temperatures.

8.2.4 Improved Catalysis

It is beyond the scope of this section to cover fully this rapidly expanding area of polymer chemistry. Indeed, Ebdon and Eastmond[14] have already published volumes reviewing recent developments in modern methods employed in the synthesis of polymeric materials and you may like to refer to these excellent sources for more information. We will simply concentrate on some developing areas of catalysis for polymer synthesis and highlight the potential benefits that they offer in lessening environmental impact.

The use of metallocene and associated single-site catalysts is a relatively recent innovation that has made it routinely possible to produce poly(l-alkene)s with a wider range of properties than has been possible hitherto using conventional polymerisation methods. Over 40 years ago, the development of Ziegler–Natta olefin polymerisation catalysts stimulated interest in the production of poly(l-alkene)s with carefully controlled stereochemistry (tacticity). Of particular practical interest was the observation that it was possible to achieve a wide range of properties (*e.g.* T_g, mechanical properties, *etc.*) from chemically identical polymers, varying only in the degree of their tacticity[15]. More recently, metallocene catalysts have been developed to allow the synthesis of atactic, isotactic or syndiotactic PP or higher poly(l-alkenes) by altering the chirality ('handedness') of the catalyst[16]. Now it is even possible to produce PP having alternating atactic and isotactic blocks (the metallocene used can be readily converted from a chiral to an achiral form and it is believed that a chiral form produces the isotacticity, while the achiral form is responsible for the atacticity). The most commonly employed metallocenes are based on zirconocene (Cp_2ZrX_2, where Cp = cyclopentadienyl and X = Cl, CH_3, *etc.*) and these tend to be used with a co-catalyst, such as methylaluminoxane (MAO, a complex oligomeric structure with molecular weights of 1000 to 15 000), to enhance the catalytic activity. A representative portion of the MAO is shown in Figure 8.5 as $[Al(CH_3)O_n$. The major difference between metallocene and conventional heterogeneous Ziegler–Natta catalysts is that the former have well-defined molecular structures and polymerisation occurs at a single site in the molecule (the transition metal atom).

The mechanism by which metallocene-mediated polymerisation takes place is still not wholly certain, but the active site is believed to be cationic in nature (see Figure 8.5).

Key Facts ⚷

● Tacticity is
believed to be
controlled by the
interaction between
π orbitals in the
transition metal
catalyst and the
last-added
monomer unit of
the polymer chain.
● Synthesising
basic polymers such
as PS with specific
'properties' such as
high temperature
stability reduces the
need for expensive
exotic polymers and
simplifies the waste
stream.

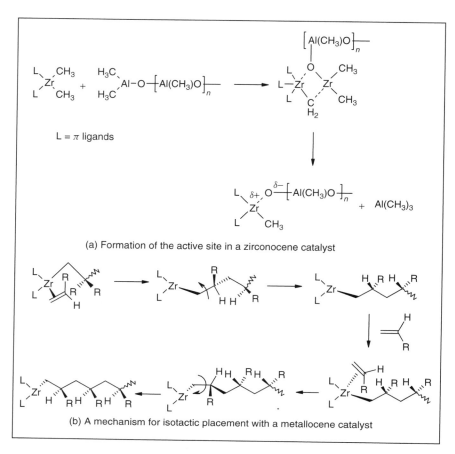

(a) Formation of the active site in a zirconocene catalyst

(b) A mechanism for isotactic placement with a metallocene catalyst

Figure 8.5 Metallocene catalysed polymerisation of 1-alkenes

After the initial formation of the active site, the mechanism by which polymerisation takes place is thought to involve initial π-complexation between the monomer and the cationic site, followed by insertion. Figure 8.5 shows the formation of an isotactic poly(l-alkene) in which the substituent R is placed on one side of the growing polymer chain. The reason for a given catalyst yielding either isotactic or syndiotactic chains is believed to involve the steric interactions between the π-ligands attached to the transition metal atom and the last inserted monomer unit of the growing polymer chain. An additional benefit arising from the use of metallocene catalysts is the narrow molecular weight distribution (MWD, see Chapter 2) and better mechanical properties that can be achieved. For instance, polydispersities range from 2 to 2.5, compared with polymers prepared using heterogenous catalysts, where the polydispersity may be 5 to 6.

By improving the control during the polymerisation process it is possible to produce a wider range of polymers with controlled architectures resulting in improved mechanical properties in many cases. For instance, the use of another single-site catalyst, such as $CpTi(OR)_3$ (where R = alkyl) has been used, in conjunction with a co-catalyst, to produce highly crystalline PS with a melting point of around 265 °C, good dimensional stability and excellent chemical and solvent resistance. These properties are in marked contrast to those exhibited by conventional atactic PS and enable the use of syndiotactic PS as a true engineering thermoplastic and allow conventional processing methods (*e.g.* injection moulding) to be used. When processed into fibres, syndiotactic PS may exhibit exceptional tensile strengths comparable with aramid fibres, a much less readily processed and consequently more expensive performance polymer. This in turn can remove the need for the more expensive material and results in the simplification of the waste stream.

A number of alternative energy sources are currently being tested to replace conventional heat sources for polymer processing, including photopolymerisation, microwave energy, electron beam and ultrasound. There are several potential advantages to using alternative polymerisation stimuli to thermal means, which may impact on either the processing or use of the polymer. For instance, the use of lower (or ambient) cure temperatures may reduce the amount of stress that builds up during network formation in a thermosetting polymer and consequently reduce the amount of micro-cracking that might occur on removing the polymer from the reaction vessel or mould. Furthermore, the degree of polymerisation, and hence T_g, may be increased when using dielectric microwave heating in comparison with thermal means (leading to a potentially higher performance polymer). However, in the present context, the primary aim of developing alternatives to thermal polymerisation is to increase the efficiency that may be achieved by using some of the techniques outlined here. Thermal cure is a relatively inefficient way of stimulating a polymerisation, particularly when processing large components, such as structural polymer composites, in an autoclave (a chamber in which the temperature and pressure may be altered to enable cure to take place, which might measure several metres or more in diameter). In such a vessel, the delivery of thermal energy to the reaction medium (the polymer matrix) is very inefficient with energy being dissipated through conduction and convection. As thermal cure (and the application of a vacuum) generally requires the generation of large quantities of electrical power, then there is a significant environmental burden associated with its use. This is particularly important when cure schedules might involve prolonged heating at elevated temperatures (often between 100 and 130 °C) for a number of hours, or days (for larger components). By increasing the efficiency of the source and targeting more effectively the delivery and penetration of the energy source to the reaction medium, it is possible to reduce the polymerisation time, which then also results in a reduced environmental impact. The following sections outline some of these alternative energy sources for polymerisation.

Key Facts

- Low temperature polymerisation not only saves energy but also produces a better product with a higher degree of polymerisation, which is less likely to form micro-cracks on cooling.
- Polymerisation can be initiated or propagated by the use of UV light, microwave energy, electron beams and ultrasound, all of which carry a lower environmental burden than thermally supported polymerisation.

Photopolymerisation

The use of ultraviolet (UV) radiation to initiate the cure of photosensitive polymers is probably the most widely developed of the alternatives to the relatively energy inefficient thermal polymerisation route and has been the subject of a review article[17]. Photo-initiators are often thermally stable, complex aromatic salts of Brønsted acids, such as diaryliodonium-, triarylsulphonium- and dialkyl-4-hydroxyphenyl sulphonium acids. Exposure to UV radiation in the region of 220–280 nm, generates the parent Brønsted acid and strong protic acids may act as powerful initiators for some polymer systems. For instance, epoxy resins polymerise *via* a series of ring opening steps following the formation of a cation with which the Brønsted acid anion is associated. Propagation occurs by successive reactions of epoxide groups with the extending and cationically terminated polymer. In the absence of impurities, termination is a very slow process, and the system may be regarded as a 'living polymer' (*i.e.* one in which propagation will continue until all the monomer is consumed).

Microwave Dielectric Heating

Many processes can be accelerated by the use of microwave energy, in much the same way that microwave cookers heat food. The microwave energy is absorbed by suitable polar ligands in the material (water in the case of food), such as hydroxyl or carbonyl bonds, C–Cl bonds, *etc.* and re-distributed through the molecule generating chemical reactions elsewhere. Microwave dielectric heating converts electromagnetic energy into thermal energy very efficiently, which, together with the volumetric nature of the heating, means that microwave dielectric heating provides an efficient and effective way of superheating

Key Facts 🔑
● e-beams generate
free radicals and
hence cross-linking.
● They accelerate
rates of
polymerisation and
hence reduce
cure times.

chemical systems. As a result, a wide range of organic reactions can be accelerated in this way for, in the liquid phase, the thermal motion results in a high frequency of molecular collisions. The use of containment allows the technique to achieve superheating of most polar solvents of $100\,^{\circ}C$ (compared with $10-30\,^{\circ}C$ by conventional thermal means), even at atmospheric pressure. This can correspond to an acceleration of chemical reactions by a factor of 10^3, a dramatic enhancement that is made even more marked by the use of microwave transparent reaction vessels (e.g. glass or PTFE) as the energy is delivered directly to the reaction mixture[18]. Microwave dielectric heating has been demonstrated to achieve chemistries that are unavailable using conventional thermal means, in a small industrial computer-controlled microwave unit for curing fibre-reinforced composites within a high pressure autoclave[19]. Most studies have concluded that cure speed is considerably increased when using microwave dielectric heating over the conventional thermal method. The use of pulsed radiation (e.g. a pulse length of 0.25 ms for a 40 W heater)[20] has also been found to increase the degree of polymerisation and improve mechanical properties, such as T_g.

Electron Beam

The use of high-energy electron beams (e-beams) can be used to induce polymerisation by the generation of free radicals within the polymer (and consequently cross-linking), potentially without the need to introduce chemically reactive functional groups (Figure 8.6).

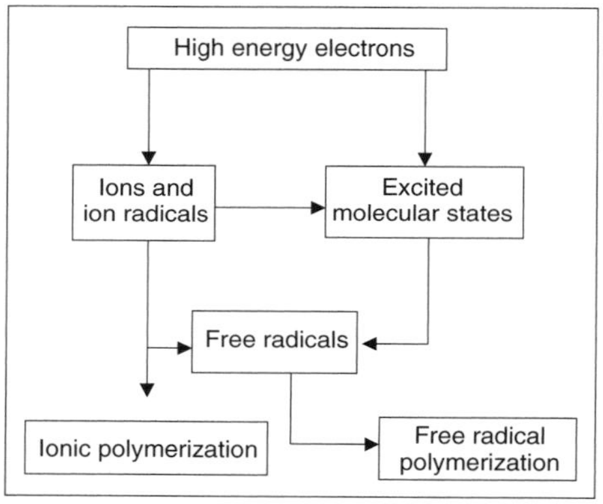

Figure 8.6 Polymerisation by e-beam initiation (High-energy electrons in e-beam curable resins and adhesives generate ionic species, free radicals, and molecules in excited states that initiate and sustain polymerisation[19])

Typical e-beam doses employed for curing polymer composites or adhesives are in the range $50-200\,kGy$[21]. The depth of penetration is proportional to the energy of the e-beam gun. For example, a 10 MeV e-beam may penetrate a component comprising composite, adhesive and substrate to a depth of 5 cm. E-beam cure can be carried out at low, near room, temperatures and can be undertaken in much shorter time-scales than the corresponding thermal cure. For instance, a combination of e-beam and X-ray curing was reported[21] to reduce the cure of a filament-wound, carbon-fibre/epoxy composite from 4 days to less than 8 h. The process has also been claimed to induce lower thermal stress in the polymer. Exposure to e-beam may lead to a number of chemical processes and both chain and step growth polymers may be initiated in this way.

The effect of e-beam exposure on resin chemistries (such as those based on free radical or methacrylated and cationic epoxy) produces different polymer structures than thermal cure, leading to concerns over qualification of 'novel' materials with consistent properties. It has also been reported that the strength of the composite fibre–matrix interface may be reduced as a result of e-beam cure, due to incompatibility between e-beam curable resins and fibre finish, although these issues are being examined in current research. It has been stated[22] that the presence of aromatic rings may dissipate the energy of the e-beam somewhat, although polymers containing aromatic rings have been demonstrated to cure rapidly in this manner[23]. Goodman and Palmese estimated[21] that e-beam processing could reduce the costs of producing aerospace composite components by between 10 and 40 %, when compared with thermal fabrication methods. Furthermore, with the use of a portable e-beam system, it is possible to cure large composite components that may not fit within an autoclave.

High Intensity Ultrasound (Sonochemical Polymerisation)

The exploration of polymer sonochemistry is gaining in momentum as it makes a wide variety of possible synthetic and modification procedures accessible. Ultrasound is typically defined (for the purposes of practical sonochemistry) as being in the range $20\,kHz-10\,MHz$, although most chemical applications use the lower end of the spectrum ($20-50\,kHz$). The reactions are promoted by cavitation, as a result of ultrasound, rather than coupling directly to the covalent bonds. Cavitation appears to act through the formation of a bubble (possibly containing solvent vapour or dissolved gases) within the mixture, which grows to around 100–200 μm (over a time-scale of around 400 μs), before collapsing rapidly (over 50 μs) to yield a 'hot spot'. The energy generated from this hot spot provides the source of activation for the chemical reactions. Cavitation may be suppressed at higher frequencies and may disappear altogether at frequencies > 2–3 MHz. In general, sonochemical activity increases proportionally with ultrasound intensity, until a threshold is reached at which too many bubbles are being produced, thus hindering the passage of sound deep into the bulk of the reactant. The presence of viscous solvents with high density, high vapour pressure or containing highly soluble gases also makes cavitation harder to achieve. Sonochemical reactions do not obey conventional Arrhenius kinetic relationships and often proceed at higher rates at lower reaction temperatures, as the bubble collapse occurs in a more intense fashion.

Key Facts
● Ultrasound generates bubbles by 'activation', and the sudden collapse of a bubble provides a high-energy 'hotspot' that promotes a chemical reaction.

To date, it appears that relatively little work has been carried out into the development of sonochemical polymerisation, although it has been examined where it may confer particular benefits (*e.g.* in radical or emulsion polymerisation where the need for initiators or emulsifiers might be avoided). It is unlikely that the technique could be applied to high tonnage production of materials without substantial technological developments. Ultrasonic degradation of polymers (*i.e.* lowering of molecular weight) also takes place in solution and may compete with sonochemical polymerisation. Degradation proceeds more effectively at higher molecular weights before approaching a limiting value (M_{lim}), below which no further degradation occurs. That means that the rate of degradation is molecular-weight dependent, as shown by Equation (8.1)[23]:

$$\ln(1/M_{1lim} - 1/M_t) = \ln(1/M_{lim} - 1/M_{init}) - k(M_{lim} - 1/cm_0)t \quad (8.1)$$

where M_{lim} is limiting molecular weight; M_t is molecular weight after sonication time t (shown in Figure 8.7 for $t = 200$ min); c is solution concentration; m_0 is monomer molecular weight and M_{init} is the initial molecular weight of polymer (for $t = 0$). The degradation proceeds faster and to lower molecular weight at lower temperatures. The process is relatively insensitive to the nature of the polymer[25,26].

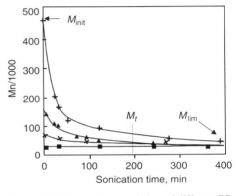

Figure 8.7 Ultrasonic degradation of different PS samples (of narrow polydispersity) in toluene solution (1 % w/v)[24] (M_n: different initial molecular weights). Reproduced with permission from J.R. Ebdon and G.C. Eastmond (eds.), '*New Methods of Polymer Synthesis*', Vol. **2**, 1995. Copyright (1995) Kluwer Academic Publishers

In this section we have concentrated on the front end of a polymer's life cycle, *i.e.* the synthesis. Continuing on a similar theme, we will now explore what improvement possibilities exist in the design of first, polymeric materials, and then the products.

8.3 DESIGNING FOR RECYCLABILITY

8.3.1 Changing the Chemistry of Degradation

The ultimate answer to the problem of polymer recycling is to build in recyclability, when the polymer is first produced and/or when products are designed. We have already discussed the factors influencing degradation in polymers (in Chapter 2) and you may remember that, whilst all polymers are susceptible to thermal, hydrolytic, oxidative or photolytic degradation to some degree or other, most synthetic polymers are immune to natural biodegradation processes. Consequently, one area that is beginning to receive more attention is the incorporation of molecular features that are inherently unstable and prone to degrade by one or more of the routes that we have already seen. Molecular design aided by molecular simulation techniques, is also being used to manufacture recyclable polymers with in-built structures that can be targeted by separation systems (*e.g.* for monomer recovery). Some examples currently being studied to build in recyclability by changing the chemistry are:

- acrylic polymers with unique solubility properties that can be recycled by the use of appropriate solvents;
- recyclable composite films that contain material that can easily be attacked to enable the breakdown of the composite (*e.g.* supermarket 'polythene' bag).

The objective is to control or modify the rate of loss of mechanical strength as the polymer degrades, by for instance, reducing the extent of cross-linking (either through the cleavage of direct inter-monomer bonds or by the disruption of hydrogen bonding) or by a direct attack on the polymer chain. However, it is important that this loss of properties happens in a predictable way over a known time-scale and it is possible to adopt several strategies (Table 8.2) to achieve this in a controlled manner.

Key Facts ✎━━━⚷
● Recyclability can be built into a polymer by incorporating ligands or filled material that can be easily attacked. Similar strategies are adopted for polymer-based drug delivery systems.

Table 8.2 Strategies to enhance the recyclability of polymer structures.

Method	Example	Feature
Add degradable side chain	Poly(ε-caprolactone)	Contains an ester linkage, which is cleaved *in vivo*
Copolymerise with degradable polymer	Poly(ε-caprolactone-*co*-DL lactic acid)	Degrades *in vivo* more rapidly than each homopolymer alone.
Add photoactive ligands	Addition, condensation, ethylenic with ketone or aldehyde ligands or residual double bonds	Light is adsorbed by the ligand and energy transferred to the polymer backbone causes scission
Add biodegradable filler	Polyester/starch; polyolefin/starch;	Starch filler undergoes degradation

Copolymerising ε-caprolactone with DL-lactic acid provides a good example of the effects of copolymerisation on stability, since the resulting product degrades more rapidly than either of the homopolymers, poly(ε-caprolactone) or poly(DL-lactic acid) (Figure 8.8). Similarly, block copolymers of polyesters and polyethers with cellulose undergo biodegradation by enzymatic attack or hydrolysis of the cellulose.

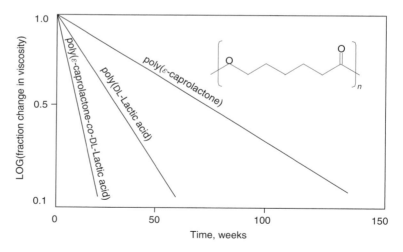

Figure 8.8 Degradation of block copolymers of poly(ε-caprolactone) or poly(DL-lactic acid)

Furthermore, photoactive ligands are now being added to thermoplastic packaging which, while capable of meeting all the usual requirements of strength, appearance and safety in use, are also capable of speedy degradation upon exposure to prolonged sunlight as litter. Many polymers are rendered photodegradable by the inclusion of functional groups, such as carbonyl or metal complex groups, which absorb light or UV radiation. There is still vast scope to improve on the technologies for production of plastics with in-built photodegradability.

Finally, an example of the use of a degradable filler to encourage biodegradation is the addition of starch to polyesters and polyolefins to reduce the environmental impact of waste plastics. A typical example is the supermarket bag, which contains up to 10 % starch. The starch degrades in the environment under the influence of moisture and bio-organisms, leaving finely divided plastic, which can be incorporated into the surroundings like any other inert material (like sand, for instance).

However, it should be borne in mind that polymer degradation represents a loss of valuable material and energy source and should only be exercised in a controlled manner, within an integrated waste management system.

Key Facts
● Supermarket plastic bags contain up to 10 % starch, which biodegrades in the environment when the bag is discarded.

8.3.2 Composites Based on Natural and Synthetic Polymers

In its broadest sense, a composite is a hybrid, *i.e.* a multi-phase material in which the interaction of two or more phases gives overall mechanical or physical properties that are significantly superior to either phase in isolation. The concept is not new of course: in nature the use of fibres to act as load bearing structures is well known, although often not recognised as such. For instance, the structure of a tree consists of long, strong cellulose fibres that are bonded together with lignin, a protein-like substance. Over many years the development of composites has progressed to the point where they are routinely being used to supplant more traditional structural materials (*e.g.* wood, glass or metals), due to the possibility of forming lightweight structures of high specific strength. However, to date most structural composites have involved the use of *e.g.* glass fibres or inorganic fillers to reinforce cheaper commodity polymer matrices, such as polyesters, or polyacrylonitrile-based fibres to reinforce advanced epoxy resin composites where higher performance is required. There has been growing interest in the use of naturally derived polymers to augment synthetic materials and one of the simplest ways in which this might be carried out is to use natural fibres to reinforce a synthetic polymer matrix. For instance, the use of modified cellulosic fibres has been examined as reinforcement for a copolymer of PS and 3-isopropenyl-α,α'-dimethylbenzyl isocyanate (Figure 8.9). The chemical modification of the cellulosic fibres enabled covalent bonding to occur between the reinforcing fibres and matrix *via* a urethane link[27].

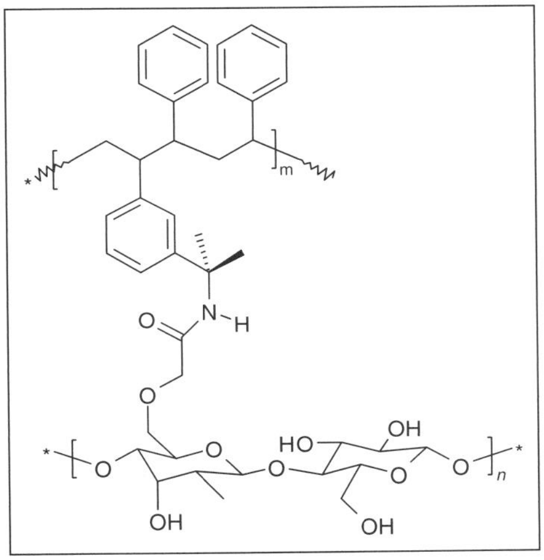

Figure 8.9 Simplified representation of the coupling of modified cellulosic fibres with poly(styrene-*co*-3-isopropenyl-α,α'-dimethylbenzyl isocyanate) *via* a urethane linkage

Lignocellulosics (whether wood or nonwood based) have also been examined as potential fillers for PP composites[27]. The advantages of this particular filler (in this case derived from the large quantity of biomass generated by the oil palm industries in Malaysia) are the lower density, greater deformability, reduced abrasion to expensive moulds and lower cost than their inorganic counterparts. The formation of composites from PP and the lignocellulosic (oil palm fruit bunch) required the addition of a suitable coupling agent, such as maleic anhydride-modified PP, to enhance the compatibility of the components. This led to considerable improvements in the mechanical properties of the composite (*e.g.* impact strength, tensile properties, flexural properties), while reducing water absorption and swelling over the PP homopolymer.

8.3.3 Building Recyclability into Polymer Products

By now you will be aware that the steps involved in separating different polymers within waste streams, in order to recycle them, are neither trivial nor cheap. A relatively simple way to overcome this problem would be to simplify the waste stream by trying to reduce the number of different polymer families in widespread use; put simply, this amounts to obtaining more properties from fewer polymers[28]. This may be achieved ideally by using existing commercial homopolymers (with improved properties arising from superior catalysis or processing) or at worst from simplified polymer blends. For instance, in a

Key Facts
● Supercritical
water (SCW), used
as a degradation
medium, converts
car tyres to oil.
● Polystyrene to the
monomer and
cellulose to a range
of hydrolysis
products.

Figure 8.10 An experimental polymeric car frame, based on a graphite–epoxy composite investigated by General Motors to achieve a 60 % saving in weight

modern mid-range car, PP now accounts for 50–60 % of all the thermoplastics on board[29] particularly in the passenger compartment. Owing to its extreme versatility, PP lends itself well to the production of single-polymer modular parts (Figure 8.10).

As we have already seen in Chapters 6 and 7, the simplification of waste streams must be done at the design stage of product development. We have also discussed there how the DFE approaches can be used to design products that are easier to dismantle and recycle, so we will not discuss this topic further here. Instead, we continue to look at the emerging technologies for polymer recycling.

8.4 NEW POLYMER RECYCLING TECHNOLOGIES

8.4.1. Use of Supercritical Fluids (SCFs)

We have seen in one of the preceding sections that supercritical fluids can be used as alternatives to organic solvents. However, of potentially greater relevance to the topic of polymer recycling, supercritical water (SCW) has been examined as a medium for the degradation of synthetic polymer waste[30]. In one study, SCW treatment at 400 °C in the absence of oxygen converted rubber tyres to oil with a yield of 44 %. PS-based ion-exchange resins have also been similarly treated (SCW at 380 °C for 1 h) to yield styrene monomer and several oxygenated arenes (*e.g.* acetophenone and benzaldehyde). Under the treatment conditions, less than 5 % of the polymer underwent decomposition. In a further development, supercritical water oxidation (SCWO) exploits the complete miscibility of organic compounds with SCW in the presence of oxygen, to yield a single fluid phase under reaction conditions. When sufficiently high temperatures are used (*e.g.* 400–600 °C), reaction rates become so rapid that near complete conversion of organic carbon may occur within a few minutes.

Although SCWO has been applied to the treatment of a variety of, often hazardous, but relatively small organic compounds (including biomass, coal oil shale, sludges, military wastes or wastewater components), the area of polymers has received little attention. One reported example is the hydrolysis of cellulose in near-critical and supercritical water in the absence of catalyst. The resulting hydrolysis of the polymer in slurry form led to complete conversion in both subcritical (350 °C, 25–40 MPa) and supercritical (400 °C, 25–40 MPa) water. The yields of hydrolysis products (cellopentaose, cellotetraose, cellabiose, glucose, fructose, glyceraldehyde, erythrose and 1,6-anhydroglucose) were higher under the more extreme conditions and the reaction was complete in very short time-scales (<50 ms), with no char formation. Some synthetic polymers have also been examined: step growth polymers such as PET, PU and some nylons (polyamides) were converted with SCW to form intermediates or monomers at near-quantitative conversions with only a few gaseous products. For example, after 12 min in SCW (at 400 °C, 40 MPa), PET was converted principally to monomer, terephthalic acid and ethylene glycol. In contrast, addition polymers, such as PE and PP, underwent much slower reaction in SCW.

8.4.2 Solvolysis

Solvolysis is a generic name for a group of depolymerisation processes, such as alcoholysis, hydrolysis, acidolysis, aminolysis and various interchange reactions that produce oligomers or monomers. Solvolytic reactions normally involve breaking of C–X bonds, where X represents heteroatoms, including O, N, P, S and halogens. Solvolytic processes are mainly applicable to thermoplastic and thermoset polymers produced by step growth polymerisation, including PET, PC, nylon and PU. A plethora of solvolysis processes are being developed and their overview is outside the scope of this chapter. Instead, we give a brief description of alcoholysis, as an example of a solvolysis process.

Polymeric materials with saponifiable groups can be converted into potential raw materials for production of PU by alcoholysis. The process of alcoholysis is analogous to hydrolysis, but employs alcohols instead of water as the cracking agents. Consequently, the mechanism of alcoholysis differs from hydrolysis and yields a mixture of the original polyol and low molecular weight urethanes with terminal alkoxy (OR) groups (if R also contains an alcohol then another urethane, bearing hydroxyl groups, is also produced). Suitable wastes can be transformed to a polyol in a batch reactor and then used directly for the production of rigid PU foams. The alcoholysis process used simultaneously with PU recycling (see Chapter 7) should also be capable of recovering or destroying CFCs in insulation foams recovered, for example, from refrigerators.

8.4.3 Mixing/Remoulding with Fresh Polymer

Blending polymer waste with virgin material at the melt stage provides a possible means of recycling certain types of polymer to produce secondary materials with properties similar to virgin material. Examples include:

- Blends of plastics waste and LDPE[31].
- PVC or PE, into which various waste plastics and impurities are blended, produces a material similar to LDPE, which is a suitable alternative to wood, metal, concrete and virgin plastics in certain applications[32].
- Photo-oxidized and stress degraded PP waste blended with virgin PP has similar properties to virgin material provided the waste content does not exceed 25 %[33,34].
- The use of polymers in concrete to make so-called polymer concretes is a well established process, but unsaturated polyester resins made from recycled PET, recovered from plastic beverage bottles, can also be used[35]. Similarly, plastics waste and scrap can be used as a replacement for sand and/or gravel in Portland Cement mixtures[36].

8.4.4 Experimental Technologies Still Under Development

Many new technologies are under development to improve recycling efficiency of polymers and reduce the costs. One, perhaps more traditional, example is solid state shear extrusion, which produces a solid product for use in lower grade applications by compressing and compacting the waste into pre-formed shapes. More recently, Stevens et al.[37] have tested dielectric microwave assisted recycling. They have demonstrated that microwave processing can drive interfacial energy absorption and adhesion in particulated thermosetting PCBs containing reactive compatibilisers with the prospect of producing materials with good mechanical properties with very low energy consumption.

Key Facts
- Solvolysis processes, such as alcoholysis, reduce a polymer to its monomer or oligomer units by a mechanism similar to hydrolysis.
- Some polymers can be recycled by mixing with virgin material or by incorporating them into materials such as cement.

Other examples involving the use of microwaves in polymer recycling process technology include:

- Dehalogenation of pesticides and PCBs with solid bases: the polarised halogen ligands readily absorb the microwave energy and the bonds are excited to such a degree that they break, releasing the halogen as a gas.
- Asphalt recycling: carbon and other conducting or semiconducting materials in the asphalt absorb the microwave energy and melt.
- Oil recovery and waste reduction: microwave absorbers in the waste heat up the mixture and vapourise off hydrocarbons and other volatile materials.
- Effects on adhesion: microwave treatment on some surfaces activates the surface to give better adhesion.
- Direct degradation: polymers with polar ligands or side changes will absorb sufficient microwave energy to disrupt the polymer backbone and induce degradation.
- Reclaiming rubber by sulphur bridging: microwaves can be used to initiate vulcanisation of rubbers.

Key Facts

- Microwaves can be used to degrade polymer materials, as well as finding applications in polymer synthesis.

8.5 BEYOND TECHNOLOGICAL SOLUTIONS

In the preceding sections we have mainly concentrated on the technical issues and the ways that polymeric materials and products can be designed in a more sustainable way. However, as we have already mentioned, technology on its own is not likely to make the world more sustainable and we have to look at the role of other agents of change. Perhaps the most important agents of change are people, and their participation is crucial if sustainable development is to be achieved.

We discussed public participation and awareness in detail in Chapter 5, but it is appropriate to restate some general principles towards the end of the book. The general public has a relatively high perception of the need for recycling as part of a sustainable future. However, the proportion of people who are actually prepared to collect the waste, separate it and take to the recycling point is much smaller. Furthermore, even fewer people are prepared to accept recycling facilities in their 'back yard', and that is particularly true for incinerators. This puts many local governments in a difficult position because of the ever-increasing mountain of waste, which we cannot continue to send to landfill indefinitely. Hence, more radical solutions are required and they call on all sections of society to contribute by reducing consumption in the first place, and that requires some changes in our life styles. So far, this is proving to be one of the greatest challenges in moving the sustainability agenda forward.

Various ways and approaches have been proposed to overcome some of the problems associated with consumer participation and many of them stress the importance of education for raising the awareness of the public. It is recognised that in many cases a lot can be achieved with little effort but also in many cases that opportunity is lost because of low levels of awareness. It is therefore important that education about sustainable development is introduced as early as possible and that is best achieved through integration into the school curriculum. Only by offering educational material sufficiently early and throughout a child's development can well-informed, future decision-makers take the necessary steps towards a sustainable future.

8.6 SUMMARY AND LEARNING OUTCOMES

We have tackled a number of different issues within this chapter examining the emergence of new technologies and how they might be applied to the subject of polymer recycling. After reading the contents of this chapter you should be able to discuss:

- The principles of 'green chemistry' and how they might be applied to the synthesis of monomers, including the use of alternative feedstocks, energy sources and molecular simulation.

- The steps being taken to improve polymer processing by reducing the use of organic solvents, using modern catalysts to produce polymers more efficiently and with tailored physical and mechanical properties, and by using more efficient energy sources to effect polymerisation and reduce production times.
- How the waste streams might be simplified by reducing our reliance on a large number of polymers (by the production of fewer polymers with controllable properties) and using the concept of designing for recycling. The introduction of specific structural features into polymers to modify their degradation characteristics and enhance their ability to be recycled is also relevant.
- How we might improve design methods in the future in order that we deliberately build new structures bearing in mind the need to recycle them at the end of their useful life.

8.7 REFERENCES AND FURTHER READING

1. Anastas, P.T. and Warner, J.C. (1998). *Green Chemistry Theory and Practice*, Oxford University Press, Oxford, p. 22.
2. Goodman, J.M. (1998). *Chemical Applications of Molecular Modelling*, Royal Society of Chemistry, Cambridge.
3. Hamerton, I., Heald, C.R. and Howlin, B.J. (1996). *Macromol. Theory Simul.*, **5**, 305–320.
4. Barton, J.M., Deazle, A.S., Hamerton, I., Howlin, B.J. and Jones, J.R. (1997). The application of molecular simulation to the rational design of new materials. 2. Prediction of the physico-mechanical properties of linear epoxy systems, *Polymer*, **38**, 4305–4310.
5. Deazle, A.S., Hamerton, I., Heald, C.R. and Howlin, B.J. (1996). Molecular modelling of high performance polymers, *Polym. Int.*, **41**, 151–157.
6. Hamerton, I., Howlin, B.J., Klewpatinond, P. and Takeda, S. (2002). Conformational studies of polycyanurates – a study of internal stress *versus* molecular structure, *Polymer*, **43**, 4599–4604
7. Gandini, A. and Belgacem, N.M. (1998). Recent advances in the elaboration of polymeric materials derived from biomass components, *Polym. Int.*, **47**, 267–276.
8. Donnelly, M.J. (1998). In vitro enzymic synthesis of polymers containing saccharides, lignins, proteins or related compounds: a review, *Polym. Int.*, **47**, 257–266.
9. Cooper, A.I. (2000). Polymer synthesis and processing using supercritical carbon dioxide, *J. Mater. Chem.*, 207–234.
10. Hartschuh, J.J., Cabiro, G.H., Dalton, M.A., Grandke, R.P., Smith, T. and Oliver, G. (1997). Emulsion and solution polymers. In *Chemistry and Technology of Water Based Inks*, ed. Laden, P., Blackie Academic and Professional, London, pp. 190–207.
11. Auchter, G. (1990). U. S. Patent 4,592,623 to BASF.
12. Kuhlkamp, A., Zimmerman, J.W. and Bütschli, L. (1967). U. S. Patent 3,345,336 to Farbwerke Hoechst.
13. Uraneck, C.A. and Goertz, R.J. (1958). U. S. Patent 2,822,354 to Philips Petroleum Co.
14. Ebdon, J.R. and Eastmond, G.C. (eds) (1995). *New Methods of Polymer Synthesis*, Vol. 2, Blackie Academic and Professional, Glasgow.
15. Natta, G. (1960). *J. Polym. Sci.*, **48**, 219.
16. Huang, J. and Rempel, G.L. (1995). *Prog. Polym. Sci.*, **20**, 459.
17. Crivello, J.V., Lam, J.H.W. (1976). New photoinitiators for cationic polymerization, *J. Polym. Sci. Polym. Chem. Symp.*, **56**, 383–395.
18. DeMeuse, M.T., Ryan, C.L., Occhiello, E., Garbassi, F. and Po, R. (1991). Chemistry and processing of polymeric materials at microwave frequencies, *Polym. News*, **16**, 262–265.
19. Boey, F.Y.C., Gosling, I. and Lye, S.W. (1993). Processing problems and solutions for an industrial automated microwave curing system for a thermoset composite, Vol. 3, Composites modelling and processing science, ICCM/9 Conference Proceedings, Madrid, 12–16 July, pp. 651–656.
20. Wei, J.B., Shidaker, T. and Hawley, M.C. (1996). Recent progress in microwave processing of polymers and composites, *Trends Polym. Sci.*, **4**, 18–24.

21. Goodman, D.L. and Palmese, G.R. (2002). Curing and bonding of composites using electron beam processing. In *Handbook of Polymer Bends and Composites*, ed. Kulshreshtha, A. and Vasile, C., RAPRA Technology, Shropshire, in press.

22. Hay, J.N., Hamerton, I., Howlin, B.J., Howgate, G.J. and O'Gara, P.M. (2001). Electron-beam cure of phenylethynyl-terminated oligomers. In *Proceedings of 'Polymer '01'*, 9–11 April, IOM Communications, Bath, pp. 288–290.

23. Ovenall, D.W. Hastings, G.W. and Allen, P.E. (1958). *J. Polym. Sci.*, **33**, 207.

24. Price, G.J. and Smith, P.F. (1993). *Eur. Polym. J.* **29**, 419.

25. Schoon, T.G. and Rieber, T. (1971). *Angew. Makromol. Chem.*, **15**, 263.

26. Schoon, T.G. and Rieber, T. (1972). *Angew. Makromol. Chem.*, **23**, 43.

27. Rozman, H.D., Lai, C.Y., Ismail, H. and Mohd Ishak, Z.A. (2000). The effect of coupling agents on the mechanical and physical properties of oil palm empty fruit bunch-polypropylene composites, *Polym. Int.*, **49**, 1273–1278.

28. Russotto, N. (1997). Plastics: using less to do more, *Mater. World*, September, pp. 529–530.

29. Seiler, E. (1995). Properties and applications of recycled polypropylene. In *Recycling and Recovery of Plastics*, ed. Brandrup, J., Bittner, M., Menges, G., Michaeli, W., Hanser, Munich, pp. 599–614.

30. Savage, P.E. (1999). Organic chemical reactions in supercritical water, *Chem. Rev.*, **99**, 603–621.

31. LaMantia, FP. (1992). Recycling of heterogeneous plastics wastes. I, *Polym. Degrad. Stab.*, **37**(2), 145–148.

32. Carrier, K. (1989). Technology for recycling post-consumer and industrial waste plastic into new, high quality products. In *Annual Technical Conference – Society of Plastics Engineers*, Society of Plastics Engineers, Brookfield Center, CT, pp. 1807–1808.

33. Valenza, A. and LaMantia, F.P. (1987). Recycling of polymer waste: Part 1 – Photo-oxidized polypropylene, *Polym. Degrad. Stab.*, **19**(2), 135–145.

34. Valenza, A. and LaMantia, F.P. (1988). Recycling of polymer waste: Part II – Stress degraded polypropylene, *Polym. Degrad. Stab.*, **20**(1), 63–73.

35. Rebeiz, K.S., Fowler, D.W. and Paul, D.R. (1993). Recycling plastics in polymer concrete for construction applications, *J. Mater. Civil Eng.*, **5**(2), 237–248.

36. Smith, B. (1986). Preliminary report on an ongoing investigation into the use of plastics, refuse and scrap as a replacement for sand and/or gravel in portland cement mixtures: To demonstrate the economic and ecological viability of recycling plastics in this manner. In *44th Annual Technical Conference – Society of Plastics Engineers*, Society of Plastics Engineers, Brookfield Center, CT, pp. 1465–1470.

37. Stevens, G.C., Hamerton, I. and Blackett, P. A feasibility study on the microwave sintering of high value waste thermosets and composites, WMR3 Dissemination event, EPSRC/DTI, Commonwealth Institute, London, 26th April, 2001.

Further Reading

Dyson, R.W. (ed.) (1998). *Specialty Polymers*, 2nd edn, Blackie Academic and Professional, Glasgow (see Chapter 8 by Hamerton, I. and Hay, J.N. for a discussion of composites).

Ebdon, J.R. (ed.) (1993). *New Methods of Polymer Synthesis*, Blackie Academic and Professional, Glasgow.

Hatakeyama, H. and Hirose, S. (1998). Environmentally compatible polymer architectures from saccharides, lignin and proteins, *Polym. Int.* **47** (1998), special edition.

Kirkwood, R.C. and Longley, A.J. (eds) (1995). *Clean Technology and the Environment*, Blackie Academic and Professional, Glasgow.

waste management in the uk

A1.1 THE EUROPEAN FRAMEWORK DIRECTIVE ON WASTE

The European regime relating to the control of waste is founded on the 1975 Framework Directive[1] on waste as amended in 1991[2]. This Directive states that its objectives include the protection of human health and the environment and that the recovery of waste should be encouraged. It provides for a system of permits for the treatment, storage and tipping of waste and provides for the application of the 'Polluter Pays' principle in respect of costs not covered by the proceeds from treating the waste. It also advocates the development of measures at national level to prevent or reduce waste by such means as the development of clean technologies and the development of products which make the smallest possible contribution to increasing waste or other pollution hazards.

The amendments in 1991 introduced a new definition of waste. This required the Commission to draw up a list of wastes falling into the categories set out in the Directive. The categories of waste are set out in Annex I of the Directive and include materials which become waste when discarded. The object of defining waste in the Directive is to introduce a common terminology and definition so as to improve the efficiency of waste management in the Community. Under the Directive all materials, substances and products may constitute waste even where they are capable of economic re-utilisation and the intention of the holder of the material is a relevant consideration in determining the question of whether the item is waste and therefore subject to the waste controls. One of the requirements of the amended Framework Directive is the protection of human health and the environment and it takes as a base a high level of environmental protection. Thus, where any doubt is raised as to whether an item is waste, this requirement will be a relevant factor.

As a Framework Directive, it has been followed by a number of daughter directives dealing with narrower areas. For example, the 1976 Directive on the disposal of polychlorinated biphenyls and polychlorinated terphenyls recognises the special hazards of these materials that are used as insulating material in transformers. Given their toxicity, the Directive prescribes particular measures to control their disposal.

A1.2 WASTE MANAGEMENT STRATEGY IN THE UK

In the UK, a significant policy document is the Waste Strategy 2000 for England and Wales[3]. This is the waste management plan required by the EU Waste Framework Directive[1,2]. It also contains the waste management plans for the Hazardous Waste Directive, the Packaging Waste Directive and the Landfill Directive. The Strategy has a 20-year life with 5-year reviews and contains specific targets for the reduction, recovery and recycling of waste.

The European definition of waste was implemented in the UK by the Waste Management Licensing Regulations 1994. These Regulations are accompanied by a lengthy and detailed circular, 'Environmental Protection Act 1990: Part II; Waste Management Licensing: The Framework Directive on Waste', which provides detailed guidance on the meaning of waste.

'Waste means any substance or object [set out in Part 11 of Schedule 4 of the Regulations] which the holder discards or intends or is required to discard'. Waste is considered to be something which poses a significantly different threat to human health or the environment partly because of the manner in which it may be disposed of and partly because the holder no longer has the same sense of obligation in relation to it. Waste, therefore, is something that falls out of the normal commercial cycle or chain of utility. This is the general test.

A recent ruling of the European Court of Justice decided that a by-product could still be waste even where it could be re-used without substantial recovery operations. This has extended the definition of waste significantly.

A1.2.1 The Waste Management Regime in the UK

The definition of waste is crucial since, once a material is classified as waste, it falls subject to the controls contained in Part II of the Environmental Protection Act 1990.

Section 33 of the Environmental Protection Act 1990 contains offences of 'knowingly causing or knowingly permitting' the deposit, *etc.*, of waste without a waste management licence.

In addition, the offence, under Section 33(1)(c), of treating, keeping or disposing of controlled waste in a manner likely to cause pollution of the environment or harm to human health can occur even if a waste management licence is in force and the conditions are being complied with. The term 'pollution of the environment' is widely defined. It includes the escape or release of matter that is capable of causing harm to man or any other living organism supported by the environment. Harm includes harm to the health of living organisms or other interference with their ecological systems. In the case of man, it includes offence to any of his senses or harm to his property. So, the smell or the sight of a landfill site can constitute an offence.

The 'Duty of Care' for Waste

The liability of producers of waste, and all others involved in the management of waste, extends beyond the moment the waste leaves their control. Valuations for insurance purposes have changed to reflect this enhanced liability. The duty of care is imposed on anyone who deals in controlled waste, (s.34 Environmental Protection Act 1990). The duty of care is designed to satisfy the European ideology on the environment that the polluter pays. The producer of waste is responsible for the proper disposal of the waste. This means that the producer must ensure it is transferred to a responsible carrier. The producer can no longer escape liability simply by passing the waste onto anyone else who could include the fly-tipper. The duty as spelled out in the EPA 1990, (s.34), is:

- to prevent the commission of one of the statutory offences;
- to prevent the escape of waste;
- on transfer to make sure it is transferred to an authorised person;
- to ensure that a written description goes with the waste so that others can comply with the duty.

So, the provisions focus on the control of waste prior to disposal and the steps to be taken on disposal. Liability after transfer will be limited to failing to take reasonable steps to detect and prevent breaches by the next person in the chain. It is likely, therefore, that a waste producer who complies with the rules of guidance on transferring waste, will be considered to have taken such reasonable steps.

This duty is imposed on all those who import, produce, carry, keep, treat or dispose of controlled waste, (s.34 Environmental Protection Act 1990). It is implemented by the Environmental Protection (Duty of Care) Regulations 1991, which are supported by a Code of Practice, 'Waste Management: the Duty of Care', and a circular issued jointly by the Department of the Environment, the Scottish Office and the Welsh Office, 'The Duty of Care'. The Circular is directed at waste disposal authorities and offers advice and interpretation of the duty of care. Local authorities are also waste producers and are also subject to the duty of care. In their capacity as waste collection authorities they collect, carry or transfer waste through direct labour organisations. Where they award contracts to the private sector they will not be waste holders. However, they will be subject to the duty of care as brokers. This means that, when an authority arranges for the transfer of waste, the correct documentation will have to be produced. The circular contains a suggested transfer form for this purpose and urges authorities to produce standard documentation. The Code of Practice provides guidance on how to discharge the duty of care imposed by s.34 Environmental Protection Act 1990. If a waste producer is taken to court for failing to comply with the duty of care, the Code can be used in evidence. It has the same status as the highway code in a traffic case. So its contents are critical to the waste industry. If waste producers fail to follow the guidelines, they are exposing themselves to prosecution.

The Code gives step by step advice and is divided into six sections:

(1) Waste producer to identify waste
Every person who handles the waste must be provided with a description of the waste so they know how to handle it.

(2) Duty to hold waste carefully
All holders of waste must keep it safely while it is under their control. They must also ensure it is in a fit state to travel. Under the new system the holder has responsibility for seeing the waste safely on its journey. The liability may not be 'cradle to grave' but is more extensive than before when it ceased once physical possession had passed.

(3) Check the transferee
Under the Controlled Waste (Registration of Carriers and Seizure of Vehicles) Regulations 1991, a comprehensive system of registration of carriers of waste was introduced from April 1, 1992. Registration may be refused if the carrier has been convicted of an offence connected with waste management and the authority think it would be undesirable.

(4) Check the transferor
The transferee of waste must check that waste is not received from a source that is apparently in breach of the duty of care. The transfer note must be properly completed and the registration of the carrier delivering the waste should be checked. This means that a carrier without proper registration should be turned away; hence the concern in the industry that the new rules would not percolate through to the numerous small firms and one-man operations engaged in carrying waste.

(5) Checking the destination of the waste
There is no specific duty on waste producers to audit the final destination of their waste. However, there is some encouragement for such a practice in the Code which states that such an audit and periodic site visits would provide evidence that an attempt had been made to prevent subsequent illegal treatment of the waste. The waste manager should have a look to see that it appears to match the description. The practice of undertaking full checks on the composition of samples of the waste is encouraged.

(6) Expert help and advice
Finally, the Code refers to the availability of advice from waste consultants and emphasises the primary responsibility of waste holders to discharge their duty of care.

All persons who import, produce, carry, keep, treat or dispose of controlled waste, and persons having control of such waste as brokers, owe a duty to take care that an offence is not committed. There is no precise definition of 'broker'. However, the Code states that a waste consultant who is directing the eventual destination of the waste may be caught by the duty. The law requires that all reasonable precautions and all due diligence have been exercised. The offence is not absolute. If all reasonable steps have been taken to prevent an escape causing damage then no prosecution will succeed. A trade practice or custom may be evidence of what is reasonable. On the other hand, a court may decide that a custom of the trade is a bad practice. So, a review of working practices may be advisable.

Special Waste

The 1991 Directive on hazardous waste prescribes more stringent rules for dealing with dangerous waste. It provides for the recording and identification of such waste when it is tipped and sets out rules for the mixing of such waste. The Directive lists in Annex I categories or generic types of hazardous waste listed according to their nature or the activity that generated them. Such waste may be liquid, sludge or solid in form. Annex I is divided into two groups. Group A lists wastes which will be hazardous if they display any of the properties in Annex III. Group B lists wastes which will be treated as hazardous if they both display any of the properties listed in Annex III and also contain any of the constituents in Annex II. Annex II lists the constituents of wastes which render them hazardous (such as beryllium compounds, chromium compounds and peroxides) when they have the properties described in Annex III (such as explosive, highly flammable, carcinogenic, corrosive, harmful). Further to these definitions a list of hazardous wastes is established in the Council Decision of 22 December 1994 and a European Waste Catalogue is established by a Commission decision for the purpose of achieving a common terminology for different types of waste.

This Directive is implemented in the UK by the Special Waste Regulations 1996. These Regulations prescribe a system of consignment notes identifiable by codes allocated by the Agencies. Pre-notification procedures are also prescribed so that information regarding the destination of special waste is made available to the Agencies. Documentation must be kept by all involved in the handling of special waste for 3 years and special waste must not be mixed. It is an offence not to comply with the Regulations except where that failure is caused by an emergency or grave danger and all reasonable steps were taken to minimise the threat to the public or the environment and compliance eventually took place as soon as reasonably practicable. The duty of care under s.34 of the Environmental Protection Act 1990 also applies to special waste.

REFERENCES

1. EEC (1975). Directive on Waste 75/442/EEC. *Offic. J.*, **L194**, 25 July 1975.
2. EEC (1991). Directive 91/156/EEC amending Directive 75/442/EEC on Waste. *Offic. J.*, **L78**, 26 March 1991.
3. *Environmental Protection Act 1990*, s.44A and *Environment Act 1995*, s.92, HMSO, London.

Definition of environmental impacts

This appendix gives an overview of the calculation procedure to estimate the contributions of environmental burdens identified in the Inventory Analysis phase to the different impact categories. The procedure is based on the problem-oriented approach[1]. All impact categories are normalised to the functional unit. The numerical values of the classification factors of some of the burdens are given in Table A2.

A2.1 ABIOTIC RESOURCE DEPLETION

Abiotic resource depletion includes depletion of nonrenewable resources, *i.e.* fossil fuels, metals and minerals. The total impact is calculated as:

$$E_1 = \sum_{k=1}^{K} \frac{B_k}{ec_{1,k}} \qquad (-) \tag{A2.1}$$

where B_k is the quantity of a resource used *per* functional unit and $ec_{1,k}$ represents total estimated world reserves of that resource.

A2.2 GLOBAL WARMING POTENTIAL

Global warming potential (GWP) is calculated as a sum of emissions of the greenhouse gases (CO_2, N_2O, CH_4 and VOCs) multiplied by their respective GWP factors, $ec_{2,k}$:

$$E_2 = \sum_{k=1}^{K} ec_{2,k} B_k \qquad (kg) \tag{A2.2}$$

where B_k represents the emission of greenhouse gas k. GWP factors, $ec_{2,k}$, for different greenhouse gases are expressed relative to the global warming potential of CO_2, which is therefore defined to be unity. The values of GWP depend on the time horizon over which the global warming effect is assessed. GWP factors for shorter times (20 and 50 years) provide an indication of the short-term effects of greenhouse gases on the climate, while GWP for longer periods (100 and 500 years) are used to predict the cumulative effects of these gases on the global climate.

A2.3 OZONE DEPLETION POTENTIAL

The ozone depletion potential (ODP) category indicates the potential of emissions of chlorofluorohydrocarbons (CFCs) and chlorinated hydrocarbons (HCs) for depleting the ozone layer and is expressed as:

$$E_3 = \sum_{k=1}^{K} ec_{3,k} B_k \qquad \text{(kg)} \tag{A2.3}$$

where B_k is the emission of ozone depleting gas k. The ODP factors $ec_{3,k}$ are expressed relative to the ozone depletion potential of CFC-11.

A2.4 ACIDIFICATION POTENTIAL

Acidification potential (AP) is based on the contributions of SO_2, NO_x, HCl, NH_3, and HF to the potential acid deposition, *i.e.* on their potential to form H^+ ions. AP is calculated

Table A2 Selected classification factors[a] for the LCA impact categories.

Burdens	Resource depletion (world reserves)	Global warming GWP 100 years (equiv. to CO_2)	Ozone depletion ODP (equiv. to CFC 11)	Acidification AP (equiv. to SO_2)	Eutrophication EP (equiv. to $PO_4{}^{3-}$)	Photochemical smog POCP (equiv. to ethylene)	Human toxicity	Aquatic toxicity ($m^3\,mg^{-1}$)
Coal reserves	8.72E + 13 tonnes							
Oil reserves	1.24E + 11 tonnes							
Gas reserves	1.09E + 14 m^3							
CO							0.012	
CO_2		1						
NO_x				0.7	0.13		0.78	
SO_2				1			1.2	
HC excl CH_4						0.416	1.7	
CH_4		11				0.007		
Aldehydes						0.443		
Chlorinated HC		400	0.5				0.98	
CFCs		5000	0.4				0.022	
Other VOC		11	0.005			0.007		
As							4700	
Hg							120	
F_2							0.48	
HCl				0.88				
HF				1.6			0.48	
NH_3				1.88			0.02	
As							1.4	1.81E + 08
Cr							0.57	9.07E + 08
Cu							0.02	1.81E + 09
Fe							0.0036	
Hg							4.7	4.54E + 11
Ni							0.057	2.99E + 08
Pb							0.79	1.81E + 09
Zn							0.0029	3.45E + 08
Fluorides							0.041	
Nitrates					0.42		0.00078	
Phosphates					1		0.00004	
Oils and greases								4.54E + 07
Ammonia					0.33		0.0017	
Chlor. solv./comp							0.29	5.44E + 07
Cyanides							0.057	
Pesticides							0.14	1.18E + 09
Phenols							0.048	5.35E + 09
COD					0.022			

[a] Classification factors are expressed in $kg\,kg^{-1}$, unless otherwise stated.

according to the formula:

$$E_4 = \sum_{k=1}^{K} ec_{4,k} B_k \qquad (\text{kg}) \qquad\qquad (A2.4)$$

where $ec_{4,k}$ represents the acidification potential of gas k expressed relative to the AP of SO_2, and B_k is its emission in kg *per* functional unit.

A2.5 EUTROPHICATION POTENTIAL

Eutrophication potential (EP) is defined as the potential to cause over-fertilisation of water and soil, which can result in increased growth of biomass. It is calculated as:

$$E_5 = \sum_{k=1}^{K} ec_{5,k} B_k \qquad (\text{kg}) \qquad\qquad (A2.5)$$

where B_k is an emission of species such as NO_x, NH_4^+, N, PO_4^{3-}, P, and COD and $ec_{5,k}$ are their respective eutrophication potentials. EP is expressed relative to PO_4^{3-}.

A2.6 PHOTOCHEMICAL OXIDANTS CREATION POTENTIAL

Photochemical oxidants creation potential (POCP), or photochemical smog, is usually expressed relative to the POCP classification factors of ethylene and is calculated as:

$$E_6 = \sum_{k=1}^{K} ec_{6,k} B_k \qquad (\text{kg}) \qquad\qquad (A2.6)$$

B_k are the emissions of different contributory species, primarily VOCs, classified into the following categories: alkanes, halogenated HCs, alcohols, ketones, esters, ethers, olefins, acetylenes, aromatics and aldehydes; $ec_{6,k}$ are their respective classification factors for photochemical oxidation formation.

A2.7 HUMAN TOXICITY POTENTIAL

Human toxicity potential (HTP) is calculated by adding the releases, which are toxic to humans, to three different media, *i.e.* air, water and soil:

$$E_7 = \sum_{k=1}^{K} ec_{7,kA} B_{kA} + \sum_{k=1}^{K} ec_{7,kW} B_{kW} + \sum_{k=1}^{K} ec_{7,kS} B_{kS} \quad (\text{kg}) \qquad\qquad (A2.7)$$

where $ec_{7,kA}$, $ec_{7,kW}$, and $ec_{7,kS}$ are human toxicological classification factors for the effects of the toxic emission to air, water and soil, respectively. B_{kA}, B_{kW} and B_{kS} represent the respective emissions of different toxic substances into the three media. The toxicological factors are calculated using the acceptable daily intake or the tolerable daily intake of the toxic substances. The human toxicological factors are still at an early stage of development so that HTP can only be taken as an indication and not as an absolute measure of the toxicity potential.

A2.8 AQUATIC TOXICITY POTENTIAL

Aquatic toxicity potential (ATP) can be calculated as:

$$E_{8A} = \sum_{k=1}^{K} ec_{8,kA} B_{kA} \qquad (\text{m}^3) \qquad\qquad (A2.8)$$

where $ec_{8,kA}$ represents the toxicity classification factors of different aquatic toxic substances and B_{kA} are their respective emissions to the aquatic ecosystems. ATP is based on the maximum tolerable concentrations of different toxic substances in water by aquatic organisms. Similar to the HTP, classification factors for ATP are still developing, so that it can only be used as an indication of potential toxicity.

REFERENCE

1. Heijungs, R. *et al.* (eds) (1992). *Environmental Life Cycle Assessment of Products: Background and Guide*, Multicopy, Leiden.

useful conversion formulae

SI Units		Traditional Units		SI Units
$J\,m^{-1}$	$\times 0.01873$	ft lb in^{-1}	$\times 53.40$	$J\,m^{-1}$
kg	$\times 9.4 \times 10^{-4}$	ton (UK long ton)	$\times 1016$	kg
$kJ\,m^{-2}$	$\times 0.4755$	ft lb in^{-2}	$\times 2.103$	$kJ\,m^{-2}$
$MN\,m^{-2}$ or MPa	$\times 144.9$	lb in^{-2}	$\times 0.0069$	$MN\,m^{-2}$ or MPa
kg	$\times 1000$	tonne (metric ton)	$\times 0.0001$	kg

notation and abbreviations

AIBN	Azodiisobutryonitrile
ASTM	American Society for Testing and Materials
BGA	German Federal Office of Health
BHET	*bis*(hydroxyethyl)terephthalate
CFC	Chlorofluorocarbon
Cp	Cyclopentadienyl
c	Solution concentration
DEG	Diethylene glycol
DP	Average degree of polymerisation
DP_0	Initial average degree of polymerisation (time = zero)
DP_t	Average degree of polymerisation after time t
DSC	Differential scanning calorimetry
DSD	Duales System Deutschland
EC	European Commission
FDA	Food and Drug Administration (USA)
FTIR	Fourier transform infrared (spectroscopy)
ISO	International Organization for Standardization
IUPAC	International Union of Pure and Applied Chemistry
K_v	Mark–Houwink constant
k_n	nth order rate constant
LCST	Lower critical solution temperature
MAO	Methylaluminoxane
MDI	4,4-Methylene-*bis*(phenyl isocyanate)
M_i	RMM of individual polymer chain
M_{init}	Initial molecular weight of polymer (t = zero)
M_{lim}	Limiting molecular weight
$\overline{M}_n$	Number average molar mass
m_0	Monomer molecular weight
M_t	Molecular weight after time, t
MSW	Municipal solid waste
$\overline{M}_v$	Viscosity average molar mass
$\overline{M}_w$	Weight average molar mass

MWD	Molecular weight distribution
N_i	Number of polymer chains (molecules) with an RMM of M_i
NO_x	Oxides of nitrogen
P_c	Critical pressure
PCB	Polychlorinated biphenyl
RMM	Relative molar mass
SAE	Society of Automotive Engineers
$scCO_2$	Supercritical carbon dioxide
SCF	Supercritical fluid
SCW	Supercritical water
SCWO	Supercritical water oxidation
SPI	Society of the Plastics Industry
TDI	Tolylene-2,4-diisocyanate
T_c	Critical temperature
T_g	Glass transition temperature
T_{gc}	Glass transition of a copolymer
T_m	Melting temperature
UCST	Upper critical solution temperature
UV	Ultraviolet (radiation)
V	Mark–Houwink constant
VCM	Vinyl chloride monomer
VCR	Video cassette recorder
V_n	Volume fraction ($n = 1, 2$)
VOC	Volatile organic component
WEEE	Waste electrical and electronic equipment
W_n	Weight fraction ($n = 1, 2$)
ΔG_{mix}	Gibbs' free energy of mixing
ΔH_{mix}	Enthalpy of mixing
ΔS_{mix}	Entropy of mixing
$[\eta]$	Limiting viscosity number (intrinsic viscosity)

Selected polymer structures and abbreviations [+ repeat units]

ABS	poly(acrylonitrile-co-butadiene-co-styrene)
EPS	Expanded polystyrene
E-PVC	Emulsion poly(vinyl chloride)
EVA	Poly(ethylene-co-vinyl acetate)
HDPE	High density polyethylene
HIPS	High impact polystyrene
HME-HDPE	High molecular weight high density polyethylene
LDPE	Low density polyethylene
LLDPE	Linear low density polyethylene
MDPE	Medium density polyethylene
M-PVC	Mass poly(vinyl chloride)
PC	Polycarbonate
PE	Polyethylene
PET	Poly(ethylene terephthalate)
PMMA	Poly(methyl methacrylate)
PP	Polypropylene
PS	Polystyrene

PTFE	Polytetrafluoroethylene
PU	Polyurethane
PVA	poly(vinyl alcohol)
PVB	Poly(vinyl butyral)
PVC	Poly(vinyl chloride)
PVP	Poly(vinyl pyrrolidinone)
SAN	Poly(styrene-co-acrylonitrile)
S-PVC	Suspension poly(vinyl chloride)
UHME-HDPE	Ultra high molecular weight high density polyethylene
VLDPE	Very low density polyethylene

Index

Note: Index entries referring to Figures are indicated by *italic* page numbers and Tables are indicated by **bold** page numbers; alphabetisation is letter-by-letter (ignoring spaces).